CONTEMPORARY ISSUES

Issues in Alcohol

by Lisa Wolff

Lucent Books, San Diego, CA

Other books in Series:

Issues in Biomedical Ethics
Issues in the Environment
Issues in the Information Age
Issues in Sports

Library of Congress Cataloging-in-Publication Data

Wolff, Lisa, 1954–
Issues in alcohol / by Lisa Wolff.
p. cm.—(Contemporary issues)
Includes bibliographical references and index.
Summary: Discusses the issues surrounding alcohol, including laws regulating the sale and use of alcohol, drunk driving, and the responsibility of advertisers for alcohol abuse.
ISBN 1-56006-447-1 (lib. : alk. paper)
1. Alcoholism—United States—Juvenile literature. 2. Drinking of alcoholic beverages—Law and legislation—United States—Juvenile literature. 3. Drunk driving—United States—Juvenile literature. [1. Alcoholism. 2. Alcohol. 3. Drunk driving.]
I. Title. II. Series.
HV5066.W65 1999
362.292'0973—dc21

98-35532
CIP
AC

P.O. Box 289011
San Diego, CA 92198-9011
Printed in the U.S.A.

Table of Contents

Foreword

> When men are brought face to face with their opponents, forced to listen and learn and mend their ideas, they cease to be children and savages and begin to live like civilized men. Then only is freedom a reality, when men may voice their opinions because they must examine their opinions.
>
> Walter Lippmann, American editor and writer

CONTROVERSY FOSTERS DEBATE. The very mention of a controversial issue prompts listeners to choose sides and offer opinion. But seeing beyond one's opinions is often difficult. As Walter Lippmann implies, true reasoning comes from the ability to appreciate and understand a multiplicity of viewpoints. This ability to assess the range of opinions is not innate; it is learned by the careful study of an issue. Those who wish to reason well, as Lippmann attests, must be willing to examine their own opinions even as they weigh the positive and negative qualities of the opinions of others.

The *Contemporary Issues* series explores controversial topics through the lens of opinion. The series addresses some of today's most debated issues and, drawing on the diversity of opinions, presents a narrative that reflects the controversy surrounding those issues. All of the quoted testimonies are taken from primary sources and represent both prominent and lesser-known persons who have argued these topics. For example, the title on biomedical ethics contains the views of physicians commenting on both sides of the physician-assisted suicide issue: Some wage a moral argument that assisted suicide allows patients to die with dignity, while others assert that assisted suicide violates the Hippocratic oath. Yet the book also includes the opinions of those who see the issue in a more personal way. The relative of a person who died by assisted suicide feels the loss of a loved one and makes a plaintive cry against it, while companions of another assisted suicide victim attest that their friend no longer wanted to endure the agony of a slow death. The profusion of quotes illustrates the range of thoughts and emotions that impinge on any debate. Displaying the range of perspectives, the series is designed to show how personal belief—whether informed by statistical evidence, religious conviction, or public opinion—shapes and complicates arguments.

Each title in the *Contemporary Issues* series discusses multiple controversies within a single field of debate. The title on environmental issues, for example, contains one chapter that asks whether the Endangered Species Act should be repealed, while another asks if Americans can afford the economic and social costs of environmentalism. Narrowing the focus of debate to a specific question, each chapter sharpens the competing perspectives and investigates the philosophies and personal convictions that inform these viewpoints.

Students researching contemporary issues will find this format particularly useful in uncovering the central controversies of topics by placing them in a moral, economic, or political context that allows the students to easily see the points of disagreement. Because of this structure, the series provides an excellent launching point for further research. By clearly defining major points of contention, the series also aids readers in critically examining the structure and source of debates. While providing a resource on which to model persuasive essays, the quoted opinions also permit students to investigate the credibility and usefulness of the evidence presented.

For students contending with current issues, the ability to assess the credibility, usefulness, and persuasiveness of the testimony as well as the factual evidence given by the quoted experts is critical not only in judging the merits of these arguments but in analyzing the students' own beliefs. By plumbing the logic of another person's opinions, readers will be better able to assess their own thinking. And this, in turn, can promote the type of introspection that leads to a conviction based on reason. Though *Contemporary Issues* offers the opportunity to shape one's own opinions in light of competing or concordant philosophies, above all, it shows readers that well-reasoned, well-intentioned arguments can be countered by opposing opinions of equal worth.

Critically examining one's own opinions as well as the opinions of others is what Walter Lippmann believes makes an individual "civilized." Developing the skill early can only aid a reader's understanding of both moral conviction and political action. For students, a facility for reasoning is indispensable. Comprehending the foundations of opinions leads the student to the heart of controversy—to a recognition of what is at stake when holding a certain viewpoint. But the goal is not detached analysis; the issues are often far too immediate for that. The *Contemporary Issues* series induces the reader not only to see the shape of a current controversy, but to engage it, to respond to it, and ultimately to find one's place within it.

Introduction

Alcohol Use in the United States

SINCE THE FIRST SETTLERS REACHED New England's shores, alcohol has been a part of American culture. The colonists brought with them the British traditions of celebrating special occasions with a toast and medicating everyday ailments with whiskey. They drank to ward off the chill in winter, to escape their troubles in hard times, and to socialize in better ones.

Future generations carried on these traditions, and settlers from other parts of the world added their own. The production of wine, liquor, and ale became a big business, and their consumption boomed. Taverns opened across the country and were favorite meeting places of young and old. Whatever could not be produced locally was imported, such as rum from the West Indies, which was drunk in abundance.

Over the centuries, drinking continued to be an important part of American life. However, recognizing the harm that abusive drinking could cause, there were several attempts to make alcohol illegal. The temperance movement succeeded in banning the sale of liquor in several states in the 1850s, but these state laws were found unconstitutional and repealed. The only national prohibition of alcohol occurred in the twentieth century and led to a huge underground economy based on bootleg, or illegally sold, liquor. Americans were not about to give up one of their favorite pastimes, even if it meant breaking the law.

Prohibition, which lasted from 1920 to 1933, banned the manufacture, transportation, and sale of alcohol throughout the country. It

A nineteenth-century print advocating temperance in America.

was by all accounts a disaster, leading to the growth of organized crime around the illegal alcohol industry. Americans, who considered the right to drink one of their basic freedoms, continued to do so, whether by making "bathtub gin" in their own homes or drinking in illegal hidden clubs. Law enforcement commonly turned a blind eye to illegal alcohol use, recognizing that even the most prominent citizens continued to drink. The law was finally repealed under public pressure.

Since then, people have been seeking ways to limit the ill effects of alcohol on society rather than prohibit its use. These include restricting its sale to adults, creating tough penalties for drunk driving, and educating the public about the harm alcohol abuse can cause. They also include programs to help people who have drinking problems.

Despite its image as a way to relax and enjoy life, alcohol is a powerful and dangerous drug. It contributes to about ninety-four thousand deaths each year in the United States. Forty percent of deaths in traffic accidents involve the use of alcohol. In 1994, 3.4 million Americans were treated for alcohol-related problems. Alcohol is involved in about half of all murders and serious assaults and a high percentage of other crimes.

Alcoholism, the addiction to alcohol, is a serious disease that affects about 8 million Americans and their families. In addition to lost jobs and broken relationships, it can lead to liver failure and death. About 20 percent of suicide victims are alcoholics.

Much of the effort to control drinking has been directed at the young. The use of alcohol and other drugs is the leading cause of death and injury among teenagers and young adults. Young drivers have the highest intoxication rates in fatal car crashes, and drinking is often a factor in youth violence. Alcohol is involved in up to two-thirds of acquaintance and date rapes among teens and college students. Using alcohol or other drugs at an early age often leads to future abuse.

At the same time, much advertising of alcoholic drinks is aimed at a young market. Sports programming, watched by a high proportion of teenage boys and young men, is heavily laced with beer commercials. The models in ads for beer, wine, and liquor tend to

Promotions for alcohol are prominent at many sporting events because young male sports fans are a major consumer group targeted by advertisers.

be young and active, and drinking and sex appeal are strongly linked. Drinking is equated with youth, beauty, fun, parties, and a carefree life.

In recent years, concerned parents whose children died in alcohol-related accidents have joined forces to focus on the dangers of drinking. They have pressured politicians to enact tougher laws and penalties for drunk driving and pushed the alcohol industry to stress responsible drinking. Drinking by minors has become a target, with bars and liquor stores facing high fines for serving to customers under twenty-one.

There has also been a strong movement to help people with alcohol abuse problems. People are coming to recognize that these problems affect not only the drinkers and their families but all of society. From 1985 to 1990, problems stemming from alcohol abuse cost the American economy an estimated $98.6 billion.

However, while many are fighting for even stronger laws to control alcohol abuse, others feel the law has gone too far. They believe that the reckless actions of a small part of society are causing the majority to lose a basic freedom. Some feel that recent laws, such as raising the drinking age to twenty-one nationwide, unfairly target the young. They view restrictions on the selling and advertising of alcohol as a violation of free enterprise, one of the principles on which this country was founded. In particular, people in the business of selling alcohol, such as manufacturers, importers, and bar and liquor store owners, feel the government is interfering unfairly in their right to make a living.

Alcohol remains a major part of American life. About two-thirds of adults in this country drink. The manufacture and sale of alcoholic beverages is an enormous industry with strong political power. Any effort at banning alcohol again would be met with forceful opposition.

In a society that values individual freedom, drawing the line between rights and responsibilities is hard. Lawmakers must balance the public's desire to make their own personal decisions about drinking with their demand to be protected against those who abuse this right. The debate that led to the failed experiment of Prohibition promises to continue into the twenty-first century.

Chapter 1

Should the Sale and Use of Alcohol Be More Strictly Controlled?

As the sun sets in New Mexico, 235 liquor store owners open their drive-up windows to serve drivers heading home from work. Long lines of cars pull up for quick, convenient service of liquor and beer.

"When I'm drunk, I want to go home, not to a convenience store where there are lights and people can smell me," commented one customer. "So I go to the drive-up because it's close to my house and I won't get hassled."[1] To others, particularly women, drive-up windows offer a safer alternative than getting out of their cars at night.

Despite the many satisfied customers, not everyone is pleased with the drive-up windows. New Mexico has the nation's highest rate of alcohol-related traffic deaths. In 1996, its 11.79 deaths per 100,000 people was 19 percent higher than the death rate of the second-highest state, Mississippi. Many people believe that the drive-up windows encourage the dangerous combination of drinking and driving and that they are partly to blame for the state's high rate of road deaths.

The state legislature is trying for the fourth year in a row to let New Mexico communities close their drive-up liquor windows. The proposed bill, HB19, would allow cities and counties to hold elections on whether to ban drive-up sales. Supporters of HB19 point out that the twenty-four states that ban drive-up liquor sales average 14 percent lower rates of alcohol-related traffic deaths than those that allow the sales. In McKinley County, the only place in New

Mexico where drive-up sales are not permitted, alcohol-related deaths have dropped steeply since a ban was imposed in 1991.

Liquor store owners do not believe a ban on window sales will prevent people from drinking and driving. They feel they are being blamed unfairly for New Mexico's highway death rate. One owner notes, "The gas stations and lounges have the same problem, whether to serve someone. And everywhere, that person who gets through is going to go back and get in his car."[2]

Easy access to alcohol through such conveniences as drive-up windows at liquor stores is often cited as a factor in drunk driving incidents.

Some owners do most of their business through the drive-up window and say that government interference will force them to shut down. However, when HB19 nearly passed in 1997, it was rejected when an amendment was added to compensate liquor store owners for their lost business.

The sale of alcohol in the United States is subject to many restrictions—some federal, some state, and others local. Wine, beer, and liquor sales are closely monitored and heavily taxed. To many, these regulations and taxes are needed to limit and help balance the harm caused by alcohol abuse. To others, they place an unfair burden on people who are trying to run a legal business in a free economy.

One of the main targets of those who favor regulation is underage drinking. Many feel that young people are not ready to handle alcohol responsibly. They have succeeded in passing laws to increase the minimum drinking age throughout the country.

The National Drinking Age

Before the mid-nineteenth century, there were almost no age restrictions on drinking in the United States. The movement to prevent teens from drinking took hold as the country became industrialized. Drinking became more dangerous as people—including the young—began operating heavy mechanical equipment, and this danger greatly increased as they began to drive.

Until recent years, it was left up to individual states to establish their legal drinking age. Some states prohibited anyone under twenty-one from buying or drinking alcohol, while others had drinking ages as low as eighteen. This created problems when teens crossed the border into a state with a low drinking age to buy liquor. States with a legal limit of twenty-one resented the low age limits in neighboring states, where their teenage residents could go to bars, then drive home drunk and endanger people in their own state.

Drinking ages in various states were raised and lowered over the years to match the political climate of the time. During the 1960s and early 1970s, there was a movement toward greater freedom for young people. As teenage boys were sent overseas to fight the war in Vietnam, an amendment to the Constitution gave eighteen-year-olds the right to vote. The slogan "Old enough to vote, old enough to fight, old enough to drink" emerged, and in the early 1970s about thirty states lowered their legal drinking age to eighteen.

However, a jump in alcohol-related car accidents and deaths led many states to reverse their decisions. The highest number of drunk driving arrests, 1.9 million, occurred in 1984, when thirty-three states had a minimum drinking age of twenty or under. Organizations to fight drunk driving sprang up and pressured politicians to enact laws that raised the drinking age to twenty-one. This eventually led to the creation of a national drinking age.

Many teens considered the new law unfair. Since they had many adult responsibilities, and received adult penalties for breaking the law, they felt these should be balanced with the right to drink. They were being treated as minors in this one area, paying the penalty for the small percentage of eighteen- to twenty-year-olds who—like some adults—could not drink responsibly.

Reprinted with permission from Mike Keefe.

Those who fought for the higher drinking age, however, had statistics on their side. Drivers under twenty-one are involved in a high proportion of serious car accidents, and many die as a result of drinking and driving. According to the Centers for Disease Control and Prevention, motor-vehicle crashes are the number-one cause of death of fifteen- to twenty-year-olds in the United States. In addition to being less experienced drivers than adults, teens tend to take more risks. In the early 1980s, before the higher drinking age was adopted nationwide, people sixteen to twenty years old accounted for 15 percent of drunk driving arrests; in 1996, this figure was down to 8 percent.

Despite these statistics, some researchers believe it is possible that raising the drinking age may actually end up increasing the number of alcohol-related accidents among teens. Authors Susan and Daniel Cohen describe this view:

> If kids can't drink legally in bars or clubs where there is at least some supervision, then they will find other settings in which they can drink, for example, riding around in their cars, or parked in some remote spot where they are less likely to get caught. . . . There is evidence to indicate that in areas where there are a lot of rigidly enforced rules against

drinking, total drinking may be less common, but there is more drunkenness and more alcohol-related accidents.[3]

It is also unclear how well the law deters teens from drinking. A 1997 study by the National Institute on Drug Abuse shows that 82 percent of high school seniors have used alcohol. It also indicates that 34 percent had been drunk in the past month. The Office of the Inspector General reports that about two-thirds of teens who drink say they are able to buy their own alcoholic beverages, suggesting that bars and liquor stores are not enforcing the law.

Still, many argue that without the higher drinking age, even more teenagers would be experimenting with alcohol and drinking more often. They believe that, combined with harsh penalties for drunk driving, laws against teenage drinking save many lives. Some feel that the penalties for selling or serving drinks to minors should be higher and that the laws should be better enforced.

In recent years there has been a shift in many states toward tougher laws and penalties for selling alcohol to minors. Sting operations, in which undercover police officers who are below the legal

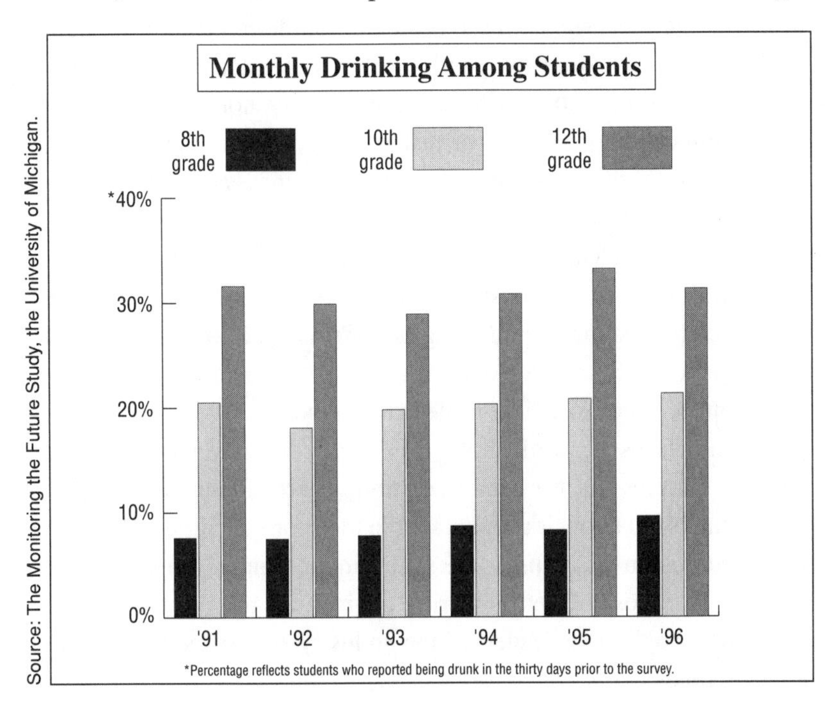

drinking age pose as bar and liquor store customers, showed how widespread the problem was. In an undercover operation in Washington, D.C., almost half of the 220 stores, restaurants, and bars tested sold alcohol to a twenty-year-old police investigator. Some did so even after checking identification that showed the customer was underage.

Restricting When and Where People Can Drink

Restrictions on drinking aren't limited to underage drinkers. Many public places ban alcoholic beverages, as do many transportation systems. Opened containers of alcohol may not be carried in cars unless in the trunk.

Places that allow drinking often restrict its hours. State and local laws control how late a bar is allowed to serve alcohol, as well as the hours liquor stores may sell it. In many states, bars and liquor stores must close by 2 A.M., though some cities have later weekend hours.

In addition, many states have "blue laws," which prohibit liquor stores from selling on Sundays. Many people argue that in addition to restricting their rights, these laws are pointless. Prohibiting people from buying liquor on Sundays, they note, doesn't keep them from buying it on Saturday to drink on Sunday, nor does it prevent them from drinking in a bar or restaurant. All such a law seems to do is keep liquor store owners from doing business and force people to plan ahead.

States often change their laws to reflect current public opinion. Kansas, for example, is considering a bill to ease its liquor laws. The proposed changes include allowing liquor store sales on Sundays and several holidays. Some people feel it will help the economies of the border counties, which lose liquor sales to neighboring states on Sundays.

Some places allow alcoholic drinks only in certain areas. In the state of Washington, for example, public sports arenas ban alcohol in the stands but allow it at clubs within the sports facilities. Because of the confusion this causes, the state Liquor Control Board is considering a proposal to let fans bring their drinks back to their seats. Terry McLaughlin, a leader of the industry group requesting the change, notes,

> Will we make additional money? Yes. Would we be involved even if we wouldn't? Yes. It is a customer service issue. I know some folks won't believe it. But this is for a person who'd like to sit down with their friend, their wife or their husband while they finish their drink.[4]

Not everyone is pleased with the idea. Nate Ford, the chairman of the Liquor Control Board, states, "I need to be convinced that's a logical thing to do, and I'm not convinced. This is an expansion of where and how liquor is consumed in places with crowds of 60,000 to 80,000 in size. Any time you are regulating a crowd that size, it is more difficult."[5]

A stronger objection was made by the director of the Central Washington Family Alliance, Sandra Swanson, who feels that alcohol should be banned in public arenas: "I was a bartender for years, and I know alcohol loosens the tongue and causes people to act in ways they otherwise wouldn't. Expanding access to hard liquor in these places makes it harder for them to be places we want to take children."[6]

Many restrictions on drinking involve places where children are likely to be. Few, if any, people would argue against a ban on alcohol in schools and day care centers. However, there is disagreement about whether places where parents are likely to bring children, such as public parks and beaches, should be alcohol-free. These issues are usually decided by local governments and often reflect the political climate of the community.

The workplace is another area where alcohol use is usually restricted. In most cases, the issues are worker productivity and providing a businesslike environment; in some, issues of safety are also involved.

Restrictions on the Job

Drinking during work hours is not encouraged in any profession. Drinking slows thinking and impairs judgment, making employees less productive and more likely to make errors. However, in certain jobs, it is typical for people to take their customers or clients to lunch and join them in drinking. Salespeople trying to establish a new account will often drink with their new client to be sociable and as a sign of goodwill.

However, in many professions, even one drink over lunch can have serious consequences. People who operate dangerous machinery could be risking their own lives as well as those of others. In New York City on August 28, 1991, a subway motorman who had been drinking before his shift derailed a train carrying hundreds of passengers, killing five and injuring more than two hundred. He had been going more than four times faster than the speed limit and was convicted of manslaughter. Airline pilots, truck drivers, bus drivers, and chauffeurs are among the many workers for whom drinking can be deadly.

Another disastrous accident involving alcohol was the crash of the *Exxon Valdez* in 1989, which caused the worst oil spill in U.S. history. The captain of the tanker, Joseph Hazelwood, had been drinking before the accident and had a history of alcohol abuse. When he ran the *Exxon Valdez* aground off the Alaskan coast, it spilled 11.2 million barrels of crude oil into the Prince William

A manufacturing plant in Detroit posts warnings against the possession of alcohol on company premises.

A worker sprays down oil-soaked rocks during clean up of the Exxon Valdez *spill. Investigators discovered that the captain of the oil tanker had been drinking before the wreck.*

Sound, leading to widespread environmental damage. Under the verdict against Exxon and Captain Hazelwood, local residents were allowed to seek $1.5 billion in compensation for their losses and $15 billion in punitive damages.

These and other well-publicized accidents prompted the U.S. Department of Transportation (DOT) to establish an alcohol testing program in 1995 for the transportation industry. The new rules covered airlines, railroads, trucking and commercial vehicles, mass transit, and boats, and applied to 7.4 million workers. They involved testing as a condition for hiring, after an accident, when alcohol use is suspected, after treatment or rehabilitation, and at random. The DOT policy also banned transportation employees from drinking alcohol within four hours of performing any duties that affect safety (eight hours for flight crews).

Testing Workers for Alcohol Use

Some companies use drug and alcohol testing of employees to be sure their judgment and reflexes are not impaired on the job. Employees who fail these tests may be suspended or fired. There is usually a provision in the contract they sign upon hiring that they will be dismissed for drinking or using drugs. Many of the tested employees are not in positions that involve a high safety risk; the companies want to be sure they will handle their jobs responsibly. For example, some banks and brokerage firms test job applicants for drugs and alcohol before they are hired. They feel that employees with drug or alcohol problems who are handling large sums of money will be more likely to steal from the company.

There are other professions in which drinking can be just as dangerous, yet workers are not commonly tested. Doctors, particularly surgeons, place their patients' lives at risk when they treat them while under the influence of alcohol. A slight error in judgment or slip of the hand during an operation can cost patients their lives. Measuring the wrong amount of medication to give someone can be lethal. For this reason, all medical personnel, including nurses and technicians, can cause serious harm by drinking during work hours.

In most cases, employees are tested for alcohol or drugs only when their behavior shows signs of impairment—or after they have caused an accident. Even in dangerous professions that risk the lives of others, testing is infrequent. Some people feel that pilots, train operators, and others responsible for the safety of their customers should be tested before each shift. Many patients about to go under the knife would feel safer knowing their surgeon wasn't drinking before the operation.

Yet for others, on-the-job testing is a violation of individual rights. They fear that if certain employees are tested for alcohol or drugs, the trend will spread to other professions. In jobs

Testing employees for alcohol may be seen as an invasion of privacy in some instances and warranted in others.

Drinking and Drug Use During Work Hours

- Percentage of workers who have an alcohol problem: 10
- Number of work-related fatalities due to alcohol or drug use: 5,000
- Lost productivity due to alcohol and drug use: $44 billion per year
- As compared to non-users, alcohol and drug abusers:

  are 2 1/2 times more likely to leave work early, take time off, or be absent

  use 3 times more sick days

  have 4 times as many on-the-job accidents

  use 8 times as many hospital days

Source: ASAP Family

where public safety is not involved, they feel that people who are able to do their work should not have to worry about losing their position because alcohol was found in their system. Many people in government and in private businesses are already tested as a condition of being hired, even when they are applying for jobs that do not put the safety of others at risk. Some feel that this testing is an invasion of their privacy—that employers don't need to know about their private habits as long as they don't interfere with their work.

There is also a question of how effective such testing is. People who know they are being tested before hiring are unlikely to drink at that time, even if they have a problem with alcohol. Random on-the-job testing is not done frequently, so many workers who are drinking or taking drugs will be missed. Some people feel that for testing to be effective, companies must invest more money in it and perform drug and alcohol tests more often.

Drivers in certain jobs that affect public safety are held to higher standards than most other drivers. For example, in California, a school bus driver was recently pulled over for speeding while driving several children to school. The police officer thought he smelled alcohol on the driver's breath and gave him a Breathalyzer test. The test results showed him to be just over the limit for most California drivers (though within the legal limit in many other states) but over twice the limit for drivers of commercial vehicles.

The driver, who claimed he had drunk a few glasses of wine the night before but nothing the morning of his arrest, was suspended from his job and faces felony charges of child endangerment. If convicted, he could get a jail sentence.

To many people, the charges are unusually harsh. The driver's lawyer said, "He had no idea any alcohol remained in his system, and would never have driven that bus if he had. He's never been stopped for drunken driving before and has a clean driving record."[7] Others, however, believe that people who are responsible for the safety of the public—especially of children—should answer to a higher code of conduct.

In addition to restrictions on when and where people can buy and drink alcohol, there are many rules that govern the alcohol industry. These cover such areas as the labeling of alcoholic beverages and the licenses needed to sell them.

Liquor Licenses and Taxes

The sale of alcohol is regulated by several agencies at the federal, state, county, and city levels. These levels of government often have conflicting laws, leading to a very confusing maze of rules for operating bars, restaurants, and liquor stores. Getting a license to sell liquor is an often difficult task that may take several months or longer.

In addition, each type of alcoholic beverage is overseen by its own industry. The Distilled Spirits Council of the United States (DISCUS) collects several billion dollars in taxes on liquor each year. The U.S. Brewers Association collects taxes on the sale of beer, and a separate industry taxes wine. For those trying to run a small business, this mix of agencies can create a bookkeeping nightmare. Some small restaurants give up on the process of filing for a liquor license and let their customers bring in their own wine and beer.

Many who apply for licenses feel that the confusion of laws—as well as the high taxes collected by the state and federal governments—interfere with their right to run a legal business. They may lose so many customers while waiting for governmental approval that their business fails. When they finally do get a license, the high

The procedure for acquiring a liquor license and the high taxes that must be paid to keep the license bothers liquor store owners who believe that the government is making it difficult to run a legitimate business.

taxes they must pay force them to charge high prices for their drinks, driving many customers away.

According to DISCUS, liquor is the highest-taxed consumer product in the country, with taxes and fees accounting for about 44 percent of the selling price of a typical bottle. Since 1980, states have passed over two hundred tax increases on liquor and many on beer and wine. Liquor tax hikes in 1985 and 1991 caused the loss of more than 87,000 jobs in the industry, according to a DISCUS report, and 155,000 jobs when the wine and beer industries are added.

To some, however, these restrictions and taxes are important in preventing too many people from opening bars and liquor stores. They argue that there are already too many, especially in poor urban neighborhoods where heavy drinking, alcohol addiction, and alcohol-related crime are serious problems. Businesses that are heavily regulated also tend to be more careful about following the law. After all they have invested, they will not want to lose their businesses because they are caught serving to minors, violating legal hours, or continuing to serve to customers who are already drunk.

The taxes on alcoholic beverages, like those on cigarettes, can be used to help cover the economic costs of alcohol abuse to the country. Government-funded hospitals treat patients injured in alcohol-related accidents as well as people with alcohol-related diseases. Federal disability insurance pays employees who lose work time because of injuries or illnesses related to alcohol abuse.

Other Restrictions on Alcohol Sales

Bars, liquor stores, and other places that sell alcohol can be fined or even lose their licenses for violating state alcohol laws. They are required to check identification showing proof of age of any customer who may be under twenty-one. (To stave off protests, many bars and liquor stores are now posting signs that anyone who appears to be under thirty-five is subject to an ID check.) Places that serve alcohol are prohibited from serving it to customers who appear to be drunk. They can also be cited for serving past closing hour or for providing liquor in bottles, which encourage heavy drinking, rather than glasses. In addition, it is illegal to continue to serve a customer who already has two drinks on the table.

While bars are often fined for breaking these laws, it usually takes many repeated violations before their licenses are revoked. Critics of irresponsible bar management would like to see laws that take away their licenses after a certain number of violations.

The many laws that govern the sale and use of alcohol are effective only if they are enforced. The laws that are followed most often tend to be those whose violation brings the highest penalties.

Penalties for Violating the Laws

Penalties for failing to enforce laws on alcohol sales and use vary widely by state. Many areas also have much stronger enforcement than others. Urban areas tend to have the harshest penalties for breaking the laws on alcohol sales. In Washington, D.C., for example, selling alcohol to a minor can incur a fine as high as $1,000 and a jail sentence of up to six months.

Because the differences between state and local laws can be confusing, some people would like to see national laws replace them. Others, however, view this as federal interference in local business.

A new area of sales that is drawing controversy is the Internet. Companies that sell wine, beer, and liquor from their websites offer easy access to alcohol to children and teens all over the country. Because this technology is new, it requires a new set of regulations that are currently being debated. In the meantime, underage drinkers have been able to buy alcohol without proof of age, and even for deliveries, and the companies that sell it have been violating state laws.

Several states, including New York and Michigan, have conducted undercover sting operations to catch companies that sell alcohol to minors over the Internet. New York's attorney general, Dennis Vacco, reported that his operation found fourteen Internet sites that deliver alcohol to customers without proof of age. Michigan has responded to the reports by imposing high fines on these companies, forbidding them to conduct business in the state, and publicizing their violations of the law. Most states are just beginning to deal with this problem. Because the Internet makes it easy to sell alcohol across state borders, many people want to see federal laws on electronic sales created and enforced.

The government plays a strong role in setting and enforcing policies for the sale and use of alcohol. To some, it oversteps its bounds by placing tight restrictions and steep taxes on legal businesses that make it hard for them to earn a profit. To others, it doesn't do enough to ensure that businesses follow the law, especially when it comes to serving underage drinkers. This group would like to see better enforcement and higher penalties for those who break the laws. While both sides agree that government has an obligation to discourage irresponsible drinking, there is widespread disagreement on how great a role it should have.

Chapter 2

Are Drunk Driving Laws Too Lenient?

In 1980 in California, Candy Lightner's thirteen-year-old daughter was killed by a drunk driver who fled the scene of the accident. When he was caught, the driver—who had three prior arrests for driving under the influence of alcohol—was allowed to plea-bargain the charge against him from murder to vehicular manslaughter. He was sentenced to only two years in prison, and the judge let him serve his time in a work camp, and later a halfway house, instead.

Lightner, outraged at how easily this killer had been released, decided to take action to prevent more drunk drivers from getting away with their crimes. She founded the organization Mothers Against Drunk Driving (MADD), which she headed until 1985. MADD now has more than 1.1 million members and has been a strong force in changing attitudes and laws about drunk driving throughout the United States.

MADD's accomplishments are impressive. Since 1980, it has helped enact more than two thousand laws against drunk driving across the country. Members worked to get the drinking age raised to twenty-one in all states. They promoted administrative license revocation laws, which allow arresting officers to take away the driver's license of anyone who fails an alcohol breath test or refuses to take one; these laws have been passed in two-thirds of all states. MADD succeeded in getting several states to lower the legal blood alcohol content limit from .10 to .08 for adults, meaning that it takes less alcohol for people to be considered unfit to drive. It also pushed for "zero tolerance" laws for drivers under twenty-one, which more

Candy Lightner is the founder of MADD, an organization that has supported tough laws against drunk drivers. MADD has been instrumental in helping pass legislation against alcohol abuse.

than a dozen states have passed. These laws prohibit drivers under the legal drinking age from having any measurable amount of alcohol in their blood system.

Judges in more than two hundred counties are ordering drunk drivers to attend victim impact panels operated by MADD. At these sessions, crash survivors tell the offenders how drunk driving has affected their lives. MADD volunteers watch court cases involving drunk drivers and report their outcome to the media and members of the community. By publicizing the cases, they put pressure on judges to hand down tougher sentences.

Most people agree that MADD and organizations like it have focused needed attention on a serious problem. The National Highway Transportation Safety Association (NHTSA) reports that about two in five Americans will be involved in an alcohol-related crash at some time in their lives if the current rate of drunk driving continues. A 1996 study conducted by MADD shows that drunk driving is the nation's most frequently committed violent crime.

However, some people think that the law has gone too far in restricting the rights of drivers and that groups like MADD have abused their influence in pressuring judges to impose maximum penalties. By publicizing drunk driving trials and staging protests against lenient judges, they encourage some judges to hand down longer sentences than they give for crimes that involve malice, or the intent to hurt or kill. A defense attorney for people arrested for driving under the influence of alcohol observes:

Though many people endorse MADD's efforts, some believe the organization has unjustly influenced judges' decisions to impose maximum penalties.

> A conviction for a first offense DUI [driving under the influence; some states use the term DWI, driving while intoxicated] charge carries with it more mandatory jail time than some felonies and a judge has sole discretion to impose up to a year in jail without explanation. The DUI laws have also been rigged so that a citizen can lose his license even if he is found not guilty in court. The laws are complex and harsh.[8]

The attorney is especially critical of a law in the state of Washington that allows the licensing department to revoke the license of anyone arrested for drunk driving whether or not a jury finds that person guilty.

Whether current laws are too harsh or not severe enough, studies show that drunk driving is a major safety issue. Each year, American highways are the scene of many thousands of alcohol-related accidents and deaths.

Drunk Driving as a Cause of Accidents and Deaths

Between 1982 and 1995, approximately 300,274 people were killed in alcohol-related traffic accidents in the United States. In 1994 alone, 11,207 drivers died in crashes, and over 46 percent of them were intoxicated. The NHTSA reports that traffic crashes are the greatest single cause of death for people ages six through twenty-eight, and almost half of these accidents are alcohol-related. In an estimated 2.2 million drunk driving crashes each year, 1.3 million innocent victims are injured or have their vehicles damaged.

Accident rates among young drivers are especially high, and crashes caused by drunk teenage drivers are often fatal. According to the California Motor Vehicle Bureau, teens are twice as likely as adults to be involved in alcohol-related fatal crashes. A 1995 Michigan study shows that more than 23 percent of drivers involved in fatal car crashes were under twenty-five. This is the main factor that led to raising the drinking age in 1984 to twenty-one in all states. In states where the age had been lower, accident rates dropped after the minimum drinking age was increased.

Many young drivers feel unfairly singled out by the law. Since they are part of a group with a high accident rate, they must not only wait until they are twenty-one to drink but must also pay higher car

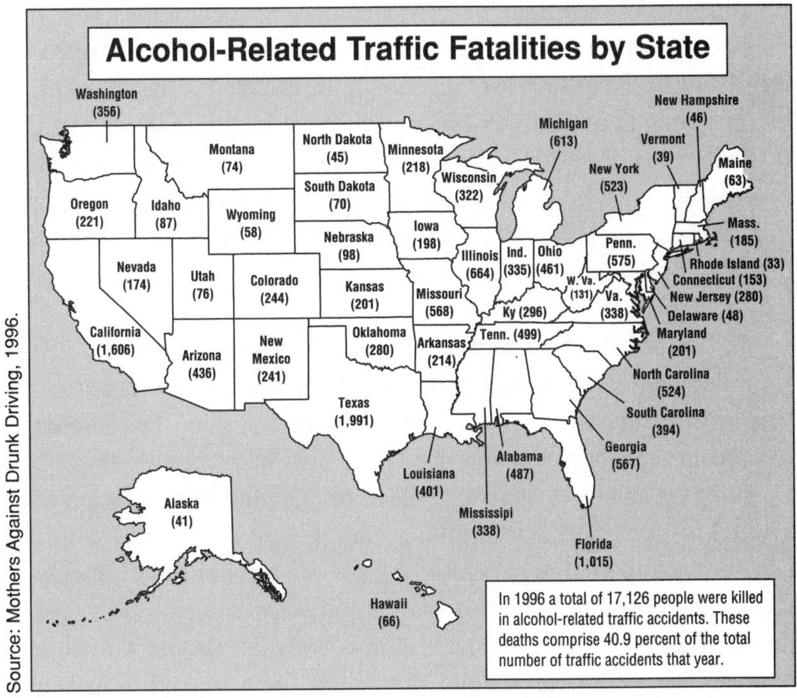

Source: Mothers Against Drunk Driving, 1996.

insurance rates than older drivers, even if their driving records are perfect. Also, while they are denied the adult privilege of drinking, the law is burdening them with adult responsibility, particularly if they are convicted of a crime. Since 1994, forty-three states have changed their laws to make it easier to prosecute minors who commit serious crimes as adults.

Whether or not certain groups of drivers are treated unfairly, the seriousness of the drunk driving problem in the United States is clear. Over half of deaths from highway accidents involve alcohol abuse. In addition to slowing the reflexes and impairing judgment, alcohol often leads to an increase in risk-taking behavior. This combination of effects can cost lives when vehicles and high speed are involved.

The amount of alcohol in a driver's system is a major factor in how dangerous that person is considered under the law. In measuring this amount, officers who pull drivers over are determining whether they will be arrested or let go.

Blood Alcohol Levels

Intoxication is measured by blood alcohol concentration (BAC). This is the measurement police use to determine whether a driver is legally under the influence of alcohol. BAC is expressed as the number of grams of alcohol per 100 milliliters of blood. In most states, it is illegal to drive with a BAC of .10 percent or more, although many states have been lowering the limit to .08 percent. According to a 1998 report by the U.S. Department of Justice, about 32 percent of fatal accidents involve a driver or pedestrian (usually a driver) with a BAC of .10 or higher.

It is impossible to determine someone's blood alcohol level just from knowing the amount that person drank. Several factors affect how quickly alcohol is processed by a person's system. The amount of food in the stomach is one factor, as food helps absorb alcohol. Drinking on an empty stomach will cause a higher concentration of alcohol in the blood than drinking after eating. A person's size is another factor; a large person will burn off alcohol more quickly than a small person, which is one reason why women tend to get drunk faster than men. Another reason is body fat: People with high amounts burn off alcohol more slowly, and women tend to have a

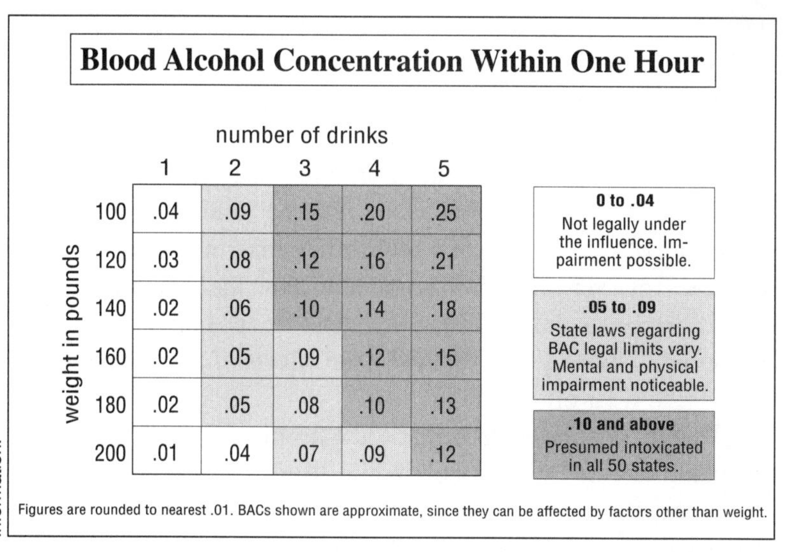

Blood Alcohol Concentration Within One Hour

weight in pounds	1 drink	2 drinks	3 drinks	4 drinks	5 drinks
100	.04	.09	.15	.20	.25
120	.03	.08	.12	.16	.21
140	.02	.06	.10	.14	.18
160	.02	.05	.09	.12	.15
180	.02	.05	.08	.10	.13
200	.01	.04	.07	.09	.12

0 to .04 Not legally under the influence. Impairment possible.

.05 to .09 State laws regarding BAC legal limits vary. Mental and physical impairment noticeable.

.10 and above Presumed intoxicated in all 50 states.

Figures are rounded to nearest .01. BACs shown are approximate, since they can be affected by factors other than weight.

Source: National Clearinghouse for Alcohol and Drug Information.

higher percentage of body fat than men. Even a person's mood can affect how quickly he or she processes alcohol.

Also, different types of drinks have different amounts of alcohol. One and one quarter ounces of 80-proof liquor has the same amount of alcohol as four ounces of wine or ten ounces of beer. Having "just one drink" is misleading if that drink is a three-ounce glass of whiskey or an eight-ounce glass of wine—enough to put many drinkers over the legal limit for driving.

Alcohol disappears from the bloodstream at an average rate of one ounce per hour. It affects the brain as long as it remains in the blood, which is why blood alcohol levels are used to determine intoxication. Police use two main methods to measure BAC: blood samples and, more often, Breathalyzers, devices that detect alcohol in exhaled breath.

The National Highway Transportation Safety Administration has done extensive studies on how drunk driving laws affect highway safety. One study examined the rates at which people charged with DWI (driving while intoxicated) refused to submit to blood alcohol tests. They found that compliance was a problem across the country, with an average of one in five drivers refusing to take the test. Refusal rates varied from a low of 2 percent in Hawaii to a high of 71 percent in Rhode Island. Rhode Island is one of only three states in which the refusal to take a BAC test cannot be admitted as evidence against an accused drunk driver in court. As a result of this study, the NHTSA recommended strong action against refusers, including criminal penalties.

Some people who have been drinking heavily think they can lower their BAC by "sobering themselves up" after drinking with a strong cup of black coffee or a cold shower, making it safe for them to drive. However, while it is possible to slow the rate at which the blood absorbs alcohol by eating, it cannot be sped up once it enters the system. The only way to reverse the effects of alcohol is to wait for them to wear off.

Another popular myth is that some people can drink a lot without being affected by alcohol. While they may not appear to be drunk, the alcohol is impairing their thinking and ability to react. Heavy drinkers who seem sober may be among the most dangerous drivers, since they are more likely to think they are in shape to drive after drinking and less likely to be pulled over until they have caused an accident.

Sobriety tests give police officers an indication of whether a driver is impaired by alcohol consumption. Even a small amount of alcohol in the bloodstream can impair a driver's attention and reaction time.

How BAC Affects Drivers

In a study of the effects of low doses of alcohol on driving skills, the National Highway Transportation Safety Administration (NHTSA) found that BACs as low as .05 affected drivers' reaction time, attention, ability to process information, vision, perception, and motor skills. It reported: "In many of these functional areas, impairment was found to appear at BACs of .02 or .03. The study concluded that no 'safe' limit of BAC, other than zero, can be placed on alcohol impairment or driving-related skills."[9] Based on its findings, the NHTSA recommended that all states consider lowering the legal limit to .08. It found that California reduced its number of alcohol-related fatalities by 12 percent in 1990 after lowering its BAC limit from .10 to .08.

Congress is currently considering a proposed law that would set a national blood alcohol concentration limit of .08. If it is approved, states would be given three years to enact the law before losing a percentage of their highway construction funds from the federal government. Critics argue that the law would violate states' rights. In an editorial, the *New York Times* countered that drunk driving is a national problem, adding:

> This [measure] has lobbyists for liquor interests trying to depict the bill as a heavy-handed assault on harmless social drinking. But a blood alcohol level of .08 is sufficient to cause unacceptable damage to a driver's reflexes, judgment and control. Moreover, the .08 level still allows for considerable consumption. An average 170-pound man, experts say, could imbibe more than four shots of hard liquor an hour—on an empty stomach—before reaching a blood alcohol concentration of .08.[10]

Some people feel that lowering the BAC limit targets responsible drinkers, when the police should be going after the heavy drinkers who pose a real danger on the road. Someone driving home from a party after a couple of drinks, they say, could fail a Breathalyzer test with a .08 limit yet be perfectly fine to drive. These critics argue that the lower limit interferes with people's freedom and is a waste of time for the police and the courts. To some, it is simply a way for the states to make more money in fines.

However, in close to one-fourth of fatal crashes involving alcohol use, the driver has a BAC of under .10. People who favor lowering the limit also point out that European countries that have even lower limits also have among the best driving records. The Scandinavian nations, for example, have a legal limit of .05 and among the safest highways in the world.

Laws in Various States

The laws and penalties for drunk driving are determined by each state. State governments control many factors, including legal blood alcohol limits, maximum sentences, and insurance rates to cover the risk of accidents.

Though all states recognize the dangers of drunk driving, some have much stricter regulation and enforcement than others. Utah, which has some of the strictest laws on the sale of alcohol, also imposes harsh penalties for drunk driving. These include taking away offenders' licenses, impounding their cars, and putting them in jail. Drunk drivers under twenty-one cannot get their licenses back until they reach that age.

Each state determines the penalties for drunk driving within its borders. Tougher laws are usually the result of public outrage over past incidents.

Many other states have been adopting tougher laws. As of late 1997, fifteen states had lowered their BAC limits to .08. Michigan recently passed four laws that increase penalties for minors and first-time offenders and make it easier for judges to seize drunk drivers' vehicles. New Mexico, a state with a high rate of alcohol-related traffic accidents, recently passed a package of laws aimed at reducing this rate. It lowered its BAC limit to .08 (with a .02 limit for minors), made it easier for police to suspend drunk drivers' licenses, made arrest of repeat offenders required, and made alcohol treatment programs part of the sentence for problem drinkers.

North Carolina, one of the states with a .08 BAC limit, was recently cited by federal safety experts as a national model in the campaign against drunk driving. Underage drinkers who are convicted of drunk driving lose their license for a year. Some offenders, such as those with a history of drunk driving, have their cars seized and turned over to local school boards to be auctioned off. The state reports that it has cut its drunk driving rate in half in recent years to the lowest in the country. To enforce its strict laws, North Carolina has set up about twenty-three thousand roadblocks, where drivers are checked for signs of intoxication, since 1993. These sobriety checkpoints are now permitted in all but ten states.

California, which has some of the toughest drunk driving laws in the country, recently introduced several bills that increase the penalties for driving under the influence of alcohol. One proposal is for drivers to lose their licenses forever after a third drunk driving conviction. The same bill would make it a crime to loan a car to someone with a revoked or suspended license. Another bill would force drunk drivers to maintain a clean record for ten years. Any conviction during this time would be considered a felony, with a mandatory jail sentence. Jim Battin, the assemblyman who introduced this bill, notes:

> We are making real progress against drunk driving. Our laws are tougher, so is enforcement. Drunk driving fatalities fell 17 percent from 1992 to 1995. Still, too many people continue to drive drunk, again and again. We fine them, we suspend their licenses, we even jail them, but sadly, they continue to drive drunk.[11]

Many other states have been taking a variety of new legal measures in recent years, which the NHTSA's Office of Program Development and Evaluation have studied for effectiveness against drunk driving. Wisconsin imposed a license suspension of three to six months for first-time offenders, resulting in a large reduction in alcohol-related crashes. Nevada takes away the license of anyone arrested for driving while intoxicated, refusing to take a Breathalyzer test, or having a BAC of .10 or more; a study showed a 12 percent drop in drunk driving accidents after the law went into effect. Oregon and Washington have vehicle plate sticker laws, allowing police officers to place a zebra sticker on the license plates of drunk drivers at the time of arrest. This alerts other traffic officers to dangerous drivers and encourages them to stop the drivers for license checks, and it has resulted in a 23 percent reduction in crashes by drivers with suspended licenses.

Much of the push for tougher sentences is a result of public outrage when a repeat offender gets only a minor penalty. In a recent Indiana case, a man with four convictions for drunk driving was sentenced to only three months in jail after his truck hit a car on an icy road, killing one of the passengers. The driver's blood alcohol level was over the legal limit and he pleaded guilty to driving while intoxicated. Because of his history of drunk driving, he was also driving with a suspended license. However, the judge set aside most of his one and one-half-year sentence because the weather had contributed to the crash. The sentencing was attended by the president of the Central Indiana branch of MADD, which is certain to keep an eye on the judge who let a serial drunk driver off so lightly.

While all rules regarding driving are left up to the states, some national organizations have recommended harsher laws. The National Research Board Committee on Alcohol, Other Drugs, and Transportation sent letters to the nation's governors urging them to take stronger measures against repeat offenders. The group's studies show that one-third of people arrested for drunk driving have prior convictions. Its letter states, "There remains a group of persistent drinking drivers who do not appear to be deterred by the threat of social disapproval or legal punishment."[12] Among its recommendations are taking away repeat offenders' vehicles, destroying their license plates, having police stake out the most serious repeat drunk drivers, and imposing harsher sentences.

Penalties for Breaking the Laws

Just as the laws against drunk driving vary widely by state, the penalties for breaking those laws are often very different, and judges are often free to decide among a wide range of sentences. However, there has been a general trend toward tougher penalties across the country. Representatives from groups like MADD have been attending the drunk driving trials of serious offenders and pressuring judges to impose the harshest possible sentence under the law.

In California, a state with some of the highest penalties, a first conviction carries a fine of up to $1,000 and may also earn a jail sentence of up to six months, followed by three to five years' probation. Judges can also order offenders to have an ignition interlocking device installed on their cars, which prevents the car from starting if the driver has a blood alcohol level over the legal limit. Fines and jail sentences increase sharply with each offense. All but five states now have laws that revoke the licenses of drivers convicted of drunk driving, though the length of time they are banned from driving varies.

Many drunk drivers whose recklessness led to fatal accidents are convicted of manslaughter (killing without that intent) and sentenced to long jail terms. Judges are taking prior drunk driving offenses into account in their sentencing to be sure repeat offenders are kept off the road.

Drunk drivers who cause fatal accidents can be charged with felony murder, the most serious category of crime. Though it is rare to find a drunk driver guilty of murder (the more common charge is manslaughter), this has happened several times in recent years. In April 1998, a drunk driver in North Carolina who killed a four-year-old girl was convicted of first-degree murder and sentenced to life in prison. He was charged under the state's felony murder rule, which allows a murder charge for a death occurring during a felony even if the death was unintentional. (In most cases, a charge of murder requires the intent to kill.) The judge's decision was influenced by the driver's five previous drunk driving convictions. Similar sentences for drunk driving were passed in Washington State in 1996 and California in 1995.

Though most people are angered by the irresponsible behavior of drunk drivers and sympathetic to the families of their victims, many feel that treating accidental killing as murder distorts the law. Killing out of negligence, they argue, is not the same as purposely taking a life and should not have the same penalties. In a 1992 California case, a drunk driver who struck five pedestrians, killing three, was charged with three counts of murder. Though she was convicted of driving under the influence, the jury was unable to decide whether the driver was guilty of murder. The judge declared a mistrial, then dismissed the murder charge instead of trying the case again, stating, "There is insufficient evidence in this case that the defendant acted with implied malice as required for a verdict of guilty of murder."[13] In other words, as terrible as the effects of the driver's irresponsible behavior were, the judge did not find proof of her intent to cause harm—a requirement for a murder verdict.

To others, though, people who get behind the wheel drunk should know that they are likely to cause harm and that a steering wheel in their hands is a deadly weapon. They see a reckless disregard for human life as being on the same scale as murder and feel that drunk drivers must take responsibility for the deaths they cause by forfeiting their freedom.

The Movement to Increase Restrictions and Penalties

In the last two decades, as groups like MADD have focused public attention on the terrible costs of irresponsible drinking, attitudes toward drinking and driving in the United States have changed. A med-

Many people believe that tougher sentencing for drunk drivers does not address the greater social problem of alcohol abuse.

ical school professor who has studied American alcohol policy observes: "Drunk driving has become a center of attention, and the killer drunk has become a mass media villain. Legislatures are passing harsher laws against driving under the influence, and politicians are promising that they will be enforced more strictly." At the same time, he notes that focusing on drunk drivers alone does not address the greater problem of alcohol abuse or a culture that makes reckless driving easy: "There is not much talk about reducing alcohol consumption, nor is the subject changed to improving automobile safety, highway design, the citing of bars, or emergency medical services."[14]

An area of major focus has been the treatment of repeat offenders. A study by the Bureau of Justice Statistics shows that more than half of all people in jail for drunk driving had prior convictions for the same offense. One in six had served at least three previous sentences for driving while intoxicated. It has become a mission for many people—particularly the parents of children who were killed by repeat offenders—to keep these reckless drivers off the road. Since so many continue to drive after their licenses are suspended, the only solution seems to be long jail sentences, combined with alcohol treatment programs, whether or not their behavior resulted in death.

Problems with Enforcing the Laws

Enacting tough drunk driving laws does not guarantee that they will be enforced. Politicians are often eager to please voters by creating strict laws and high penalties for their violation, but they often fail to come up with the funding to give these laws much power. Without enough traffic officers patrolling the highways, many drunk drivers will escape prosecution until they cause a serious accident. And many who have caused accidents remain on the road, driving with suspended licenses.

Dr. Philip Brewer, the medical director of the adult emergency department at Yale–New Haven Hospital, reports that most drunk drivers brought in for treatment are not punished for their crimes. He observes that when police officers send injured drunk drivers to the hospital, they usually don't take the time to charge them with their offenses. Most officers estimate that it takes about four hours to process a DUI case, and many drunk driving accidents occur in the middle of the night, often toward the end of an officer's shift. Many police officers hope that the doctors will take care of the reporting. However, doctors who treat the drivers often fail to report them to the Department of Motor Vehicles, nor are they required by law to do so. Dr. Brewer describes the extent of the problem: "There have been studies out of various states that have shown pretty much the same thing. One study from San Antonio [Texas] had 249 drivers in crashes brought to emergency with an average alcohol level of 1.9—that's more than double the legal limit. Of those 249, only 21 were arrested for D.U.I."[15] He recommends that doctors take more responsibility by diagnosing drunk drivers with alcohol dependency, a condition that damages their driving ability, and reporting them to the Department of Motor Vehicles.

In many states, public funding for law enforcement on the highways is not considered a top priority. People are more concerned with other crime issues that frighten them more, so state and local governments shift their resources from road safety to those areas. As George W. Black, a member of the National Transportation Safety Board, notes:

> One of the problems that traffic law enforcement has at the local level is that people are far more afraid of being murdered in their bedroom at night than being killed in a car.

> You are far more likely to be killed or injured in a car than from someone breaking into your house. But the chief of police feels compelled by elected officials to put more and more emphasis on anti-burglary and neighborhood policing, all of those touchy-feely words. Where do they get these resources? They get them out of traffic squads.[16]

In Black's view, politicians should be more responsible in informing the public on crime statistics to let them know where law enforcement funds are needed most.

The dangers of drunk driving are well known, and everyone would like to see dangerous drivers off the road before they injure or kill. While no one argues for the right to drive drunk, many critics of the tough new laws and penalties feel they are going too far. Some fear that lowering the legal BAC limit will turn people who are capable of driving responsibly after a couple of drinks into criminals. Even more controversial are laws and sentences that punish drunk drivers who cause fatal accidents as severely as murderers. Under these laws, the irresponsible act of getting behind the wheel after drinking heavily is the same as having the intent to kill. Many people believe that it is unfair and a distortion of the law to equate drunk driving—even when it results in a death—with murder. Critics of the recent move toward harsh laws and sentences would like to see more focus on education, prevention, and treatment than severe punishment of drunk drivers, many of whom suffer from alcohol addiction.

Chapter 3

Should the Law Punish Problem Drinkers?

IN APRIL 1992, TWENTY-FIVE-YEAR-OLD Michael Newbury turned himself in to the Maine police. He had awakened that morning to find himself drunk and drenched in blood, the dead body of his girlfriend beside him. Eight months later, he pleaded guilty to manslaughter at his trial, explaining, "I was in an alcoholic blackout. I don't remember anything."[17]

This was not Newbury's first alcohol-related crime. Five years earlier he had been convicted of manslaughter for driving drunk and causing an accident that killed his eighteen-year-old passenger. Yet after a short sentence, his alcoholism went untreated, and his dangerous behavior continued until he caused another violent death.

Unfortunately, this tragedy is far from unique. Each year many thousands of people commit serious crimes under the influence of alcohol, including murder, rape, and robbery. Some claim they don't even remember the act—that they were in a blackout at the time, a period during which they weren't aware of what was happening. In Maine, police, prosecutors, and judges report that 80 to 90 percent of criminals were drinking or drunk at the time they stole, raped, or killed. Meg Elam, the Cumberland County deputy district attorney, notes, "It's frustrating that so much attention is focused on illegal drugs when it's alcohol that's present in 90 percent of the cases we handle. Everybody worries about street drugs, yet alcohol is the biggest problem we face in this state. It's rare that the victim, defendant or both aren't drinking."[18]

Drunk driving is the most obvious consequence of alcohol abuse, since it leads to so many major injuries and violent deaths, often of innocent victims. It receives the most attention and is, in itself, a crime.

People who drink irresponsibly can cause many other kinds of harm to themselves and to others. Heavy drinking often leads to violent behavior, particularly in men. Parents with drinking problems often abuse or neglect their children. Women who drink heavily while pregnant may cause serious harm to their unborn children. Over time, alcohol abuse can result in severe liver damage, costing society in terms of both lost productivity and high hospital bills.

The law seldom steps into these areas. Police and the courts tend to treat abusive drinking as a private matter unless it involves driving or results in serious injury or death.

In some areas, this is beginning to change. Recent domestic violence laws make threatening behavior toward a partner—often linked to drinking—a crime even in the absence of physical abuse. Endangering the welfare of a child, even if it does not involve abuse, is against the law, and many parents with drinking problems who neglect their children's care face penalties. This leads to a debate on how much right the authorities should have to intervene in personal relationships.

Alcohol and Violent Behavior

Alcohol affects people unpredictably. There are no reliable rules about how many drinks will turn a happy, relaxed drinker into an angry, abusive drunk. Many factors affect how a person's body processes alcohol, which in turn affects that person's behavior and state of mind.

Studies have shown that alcohol affects a hormone called LH, which stimulates brain cells that play a role in regulating aggressive behavior. At the same time that alcohol depresses many body functions, it can trigger angry and violent behavior in some people. A doctor who has examined the problem notes: "It is commonly believed that drinking enhances courage and alleviates fear and apprehension in the face of danger. Although alcohol may have a salutory effect in promoting courage and aggression in certain life-threatening situations, in most instances, alcohol-related aggression is destructive to the aggressor as well as the victims."[19]

The effects of alcohol are unpredictable. Some drinkers become lethargic while others turn violent.

Aggressive behavior from drinking occurs much more often in men than in women, and women are frequently its victims. The same hormonal change that stimulates aggression increases sexual desire, and the combination sometimes leads to sexual violence, including rape. Many violent sexual attacks are done under the influence of alcohol, and many attackers use alcohol as an excuse for their behavior.

A report by the Higher Education Center for Alcohol and Other Drug Prevention notes that several studies estimate that between 50 and 80 percent of violence on college campuses is alcohol-related. In a study of female victims of sexual attacks in college, the women reported that 68 percent of their attackers had been drinking at the time. The Higher Education Center's report states,

> Alcohol contributed to violence in multiple ways, chiefly by increasing aggression, particularly when the blood-alcohol level rises rapidly (such as with binge drinking), decreasing capacity for conflict resolution, and decreasing inhibitions. In addition, many students believe intoxication excuses inappropriate and violent behavior.[20]

A recent focus on domestic violence and child abuse is bringing more attention to the destructive role alcohol often plays in relationships. One report estimates that between 25 and 50 percent of all incidents of wife abuse involve drinking. Though there have been few studies to show how important a factor alcohol is in their attacks, it is clear that heavy drinking increases irresponsible behavior that often leads to injury.

The Strong Link to Serious Crime

Alcohol is often a factor in murder. A study by the Minnesota Institute of Public Health found that, in that state, drinking was involved in nearly half of all homicides: "While there is no conclusive evidence that alcohol causes violence, it is clearly a strong factor influencing violence."[21] The Center for Substance Abuse Prevention reports that alcohol is a key factor in up to 68 percent of manslaughter cases, or killings that were not planned in advance.

A study by the National Institute of Justice found alcohol to be a major factor in a large percentage of violent crimes, though it stopped short of naming alcohol abuse as the cause of violence. Researchers noted the possibility that people first decide to commit a crime, then drink to get the courage to follow through or control their fears. Without the alcohol, however, it is likely that fewer crimes would be committed.

In a survey by the U.S. Bureau of Justice, more than half of convicted prisoners admitted to using alcohol or drugs at the time they committed their crime. Twelve percent said they used both alcohol and drugs, and 29 percent used alcohol alone. Among violent offenders, 47 percent reported using alcohol before committing their crime. The rate was even higher among those jailed for homicide: Almost two-thirds said they had been drinking at the time they killed. One in five of the inmates in the study described himself as an alcoholic.

A 1998 report by the Department of Justice shows that almost four in ten violent crimes involve the use of alcohol. It states that about 3 million violent crimes occur each year in which the victims report the attacker was drinking.

While most people agree that drunk drivers should be arrested before they have the chance to cause an accident, few would have the police arresting people in other situations simply for being drunk. However, many people feel that those who committed previous crimes while under the influence of alcohol should be prevented from repeating their behavior. There have been proposals to ban released prisoners or those on parole who committed alcohol-related violence from drinking. Some states have enacted laws that place special conditions on those who commit alcohol-related crimes. In Maine, staying away from alcohol is the most common probation condition for criminals. The *Portland Press Herald* reports that in felony cases, judges order 80 percent of criminals not to drink and to seek alcohol treatment. Still, most solutions are impractical. Unless an offender is being watched twenty-four hours a day, it is likely he or she will continue drinking.

People who live with a partner who becomes violent while drinking do have some protection under the law. They can use the person's past violent or threatening behavior as a reason to keep that person away.

Drinking and Domestic Violence

Studies show that, in the United States, 50 to 60 percent of physical attacks in the home involve alcohol use. One-third to one-half of all batterers are reported to be problem drinkers. A social worker who counsels batterers reports, "Alcohol aggravates issues related to domestic violence. What would otherwise be a minor dispute gets escalated because one or more people involved in the fight are drunk."[22]

Men with drinking problems often seem to lead responsible lives during the day. They may do well on the job, get along with their coworkers, and keep their problems hidden until they get home in the evening and start drinking. Then, alcohol will seem to change them into another person who is angry and abusive toward his family. As a district attorney notes, "Virtually every domestic violence case we see, there's alcohol involved. The victim will often tell us, 'He's really a nice guy when he's not drinking.'"[23]

Victims of violence in the home, who are usually women, can get restraining orders to protect them from further attack. The partners of physically abusive men can file a police report, then get a court order of protection to keep the abuser away from them. However, these orders address the abusive acts rather than the drinking that contributed to them. Some supporters of battered women feel that violent men who abuse alcohol should have to get treatment before they can reunite with their families.

Over half of the cases of domestic abuse in America involve alcohol consumption. Alcohol can often escalate a verbal argument into a physical conflict.

Judges sometimes include an alcohol treatment program as part of the sentence for people convicted of alcohol-related crimes. Those who fail to complete the program may have their jail sentences increased. Many people who see alcohol addiction as a lifelong problem believe that offenders should be required to receive ongoing treatment. To others, however, this amounts to lifetime parole, something even the most dangerous criminals seldom receive.

Should Treatment for Alcoholism Be Required?

Alcohol addiction is most commonly know as alcoholism. Alcoholism was once seen as a moral weakness that could be conquered through self-control. Today, most experts consider it a disease. The American Medical Association defines alcoholism as "an illness characterized by a significant impairment that is directly associated with persistent and excessive use of alcohol."[24] The National Council on Alcoholism and Drug Dependence states, "The disease is often progressive and fatal. It is characterized by impaired control over drinking, preoccupation with the drug alcohol, use of alcohol despite adverse consequences, and distortions in thinking, most notably denial."[25]

According to the National Institute of Alcohol Abuse and Alcoholism (NIAAA), about 13.8 million Americans have problems with drinking, of which 8.1 million are alcoholic. Getting accurate statistics is nearly impossible, and many people believe the number of alcoholics is higher. Alcoholics tend to deny their addiction, claiming that they can stop drinking whenever they wish. Because drinking is an accepted part of American society, the lines between social drinking, alcohol abuse, and alcoholism are not clearly drawn. This helps many people hide a serious, and often dangerous, illness.

The cost to society of alcoholism is tremendous. In addition to crimes committed by people who are drunk, according to the NIAAA, nearly one-fourth of all people admitted to general hospitals have alcohol problems, and the health care costs of untreated alcoholics are 100 percent higher than those of the general population. As with drunk driving, much of this cost is picked up by the general public through taxes and high health insurance rates.

To many, it is unfair for society to have to pick up the tab for problem drinkers. They believe that people who drink irresponsibly should have higher insurance rates than others to cover the costs of their abuse. Some insurance plans have already moved in this direction, charging higher rates for smokers than for nonsmokers to cover their higher risk of disease. Many plans refuse to enroll people who already have serious medical problems or are at high risk of developing them, while some instead charge these people higher rates to join.

To others, this is punishing people for having a disease. They argue that treatment, not punishment, is the appropriate solution. In their view, helping sick people who cannot help themselves is society's responsibility. However, when a disease is caused by an addiction, Americans are deeply divided on how responsible the victim is for his or her illness.

Are Addicted Drinkers Responsible for Their Health Problems?

Among the many costs to society of alcohol abuse is the cost of medical treatment for people with alcohol-related illness, along with their lost productivity when they are unable to work. According to the National Institute of Alcohol Abuse and Alcoholism, almost one-quarter of all people admitted to hospitals have alcohol problems that affect their health. The NIAAA reports that alcohol can harm nearly every organ and system in the body.

Heavy drinking is the most common cause of liver disease. Large amounts of alcohol cause the liver, the body's largest internal organ, to become inflamed. The liver is the main organ involved in the body's breakdown and elimination of the toxins in alcohol, so people who continue drinking after it has been damaged become seriously ill and often die. Liver disease from alcohol abuse is the fourth leading cause of death of males in American cities. Each year, about 11 million people in the United States die of the most serious form, cirrhosis of the liver. Since there is no cure, treatment is long-term and expensive.

Other disorders of the digestive system also can result from alcohol abuse. Alcohol prevents vitamins from being absorbed and transported from the intestines, causing nutritional imbalances. It can irritate the stomach lining and may cause ulcers or inflame existing ones. Functioning of the pancreas can also be impaired by heavy drinking.

Alcohol abuse can also affect the heart, and heavy drinking may lead to severe heart disease. Steady drinking of large amounts can damage the heart muscle and cause abnormal contractions that may lead to heart failure. It has also been linked to certain types of cancer.

Taxpayers cover much of the cost of alcohol-related disease through publicly funded insurance policies like Medicaid. Under the law, people who are poor enough to receive public health benefits cannot be denied free treatment even if their illness is a result of abusive drinking. Another issue is that everyone's health insurance rates are driven up by the high medical costs of alcohol abuse. Private insurance companies have already started charging higher rates for people with another addiction that leads to costly illness: smoking. Many people would like to see this extended to abusive drinkers. They feel it is unfair to charge the general public for the irresponsible behavior of others. Some believe that people who continue to drink after damaging their health should not receive financial support for their health care.

In a recent (and not uncommon) case, a woman in her forties ran up about $100,000 in medical bills for her alcohol-related problems, most of which she was unable to pay. She had started drinking in her twenties, and her abuse of alcohol increased steadily over the years. Like many alcoholics, she often substituted alcohol for food, leading to lack of important nutrients and weight loss. Her esophagus and pancreas were damaged from heavy drinking, her blood stopped clotting properly, and she ended up in the hospital fifteen times over a period of five years. Each two-week hospital stay cost about $10,000, and the woman had no health insurance. Most of the costs were picked up by the state.

A particularly controversial area in the medical treatment of alcohol abusers is liver transplants. The only hope for patients who are dying of liver disease is a new liver. However, there is a serious shortage of donor livers. Many people feel that those who abuse alcohol should be the last to receive these livers, since they were responsible for destroying their own. This drew a lot of attention in 1995 when baseball legend Mickey Mantle, who had drunk heavily for many years, was dying of liver disease. He received a new liver within forty-eight hours of being placed on a waiting list, while many patients—some whose disease is not caused by abusing alcohol—must wait years. This led some critics to push for changes in hospital policy.

On the other hand, to many who view alcoholism as a disease, denying medical treatment to people with an alcohol addiction is blaming the victims. They argue that many people who contribute to their own illness do so because they can't stop their harmful behavior. Denying them the same rights as people with other illnesses, in their view, is unfair punishment.

People often have the harshest opinion of addicts who pass their illnesses on to their children. While most feel that adults addicted to alcohol or other drugs have some choice in using the addictive substance, their children are innocent victims.

Drinking During Pregnancy

The link between drinking and birth defects has been recognized for some time. However, the seriousness of the consequences has only recently come to light.

Many children born to mothers who drank heavily while pregnant have a disorder called fetal alcohol syndrome (FAS). More than

Controversy surrounded the liver transplant operation of Mickey Mantle, a legendary sports figure and heavy drinker.

five thousand babies are born in the United States each year with this disorder, and nearly fifty thousand are born with a less severe form called fetal alcohol effects (FAE).

According to the *Journal of the American Medical Association,* fetal alcohol syndrome is the leading known cause of mental retardation. It causes a group of physical and mental defects that include poor growth, low intelligence, learning disabilities, deformities of the face and head, and problems with organ function. Most children with FAS are placed in foster homes or adopted by age five, and about 60 percent die by the age of ten.

In addition to the terrible cost to children born with FAS, the cost to society is high. The average treatment cost for each child over a lifetime is $1.4 million. Alcohol-related birth defects are estimated to cost American taxpayers $321 million each year.

The National Organization on Fetal Alcohol Syndrome reports that at least one in five pregnant women uses alcohol and/or other drugs during pregnancy. Alcohol produces the most serious side effects on the unborn child, and the producers of alcoholic drinks are required to put warning labels on their products. However, awareness of FAS remains low, and many pregnant women don't take the warnings seriously.

Some people think that drinking during pregnancy is child endangerment and should be considered a crime. They point to the recent arrest in some states of women who use other drugs, such as cocaine, while pregnant. Their sentences are based not only on illegal drug use but on endangering the life of a minor. Those who propose that women who use drugs and alcohol during pregnancy be treated as criminals often feel this is the only way to get them to stop putting new lives in danger.

To others, however, these women are being punished for having a disease. They feel that doctors should do more to explain the dangers of drinking to pregnant women and encourage them to get help. Many believe that punishment does little to stop crime in the face of addiction, which causes people to take serious risks to their own health and safety as well as that of others.

Also, it is easier—and more socially acceptable—to force drunk drivers with alcohol problems off the road than to force pregnant women to take care of their and their future children's health. American society and its legal system are reluctant to interfere with a woman's rights when it comes to having children. The courts have never been comfortable making such personal decisions.

While few people want to see all pregnant women who drink go to jail, many feel that those with drinking problems should be forced into treatment. They believe it is society's responsibility to care for its children when their parents cannot. Some propose jailing women who refuse or drop out of treatment during pregnancy. They argue that this unwillingness to accept help means the women are abandoning their responsibility and giving up their rights as caretakers of their children.

Alcohol, Child Abuse and Neglect

Even when children escape disease caused by their mother's drinking during pregnancy, they often suffer abuse or neglect from alcoholic parents. A study in New York City showed that in 64 percent of all reported child abuse and neglect cases, one or both parents abused alcohol. Parents with alcohol problems are often poorly

Reprinted with permission from Ed Gamble.

equipped to raise children. They may lash out at them when they are drunk, causing physical as well as emotional harm.

Child endangerment is a crime, though it is seldom prosecuted until serious damage—and sometimes death—has been caused. Well-publicized cases of serious child abuse, in which the parents drank, took drugs, or both, have focused attention on the problem in recent years.

While not all parents with drinking problems are violent, their children often suffer from neglect. Parents who are unable to care for themselves often fail to provide their children's basic needs. Parents caught up in addiction may neglect to feed their children regularly or see to their health needs. Those children may not get regular medical or dental checkups, vaccinations, or treatment for illnesses. They may not be warmly dressed in cold weather and may miss school because parents fail to supervise their attendance.

A woman who lost custody of her two children, ages one and two, recalls how she locked them in their room from morning until night while she drank: "It was pretty much like they were in a jail cell. Nobody would even treat a dog the way I treated them. . . . I have a lot of guilt and shame. These two children were depending on me. It's a real sickening feeling to know what I put them through." However, she also notes, "Some people think that because a mother drinks she doesn't care about her children. But that's not true. This disease takes control of your life and you don't have any choices. The bottle makes the choices."[26] The state removed the neglected children from the home after neighbors repeatedly called the Department of Health and Human Services to report their constant crying.

When abused or neglected children are removed from the home, the public must pay for their care through taxes. They must also fund treatment programs for parents with an alcohol addiction. In Maine, offering counseling, welfare, and treatment to families suffering from alcohol abuse costs about $175 million a year in state taxes.

Some people feel that parents who cannot care for their children responsibly should lose custody permanently. However, the American legal system leans heavily toward keeping families together. Throughout history, judges in the United States have been reluctant to separate children from their parents—especially their mothers—except in cases of extreme abuse. When children are removed from a dangerous home, it is usually a temporary measure until the parents have received some sort of treatment for their problems. In addition to detoxification, or supervised withdrawal from alcohol, this may involve group and individual counseling sessions and parenting classes.

Even when abused children have been placed in foster homes for a long period and their foster parents want to adopt them, judges often side with the birth mothers despite their history of alcohol abuse. To many family court judges, this is in the best interest of the child. However, many critics—especially foster parents who wish to rescue these children from an abusive home—find these decisions irresponsible. They feel that the judge's lack of responsibility compounds the mother's.

Children of Alcoholics

While there are laws to stop parents from abusing or neglecting children, those growing up in alcoholic homes often suffer in ways that are not addressed by the legal system. An estimated 15 to 17 million children grow up in the United States with at least one alcoholic parent, and this often seriously affects their lives. A report in the *Juvenile and Family Court Journal* notes, "Children of alcoholics (COAs)

Alcoholism is detrimental to the physical health of the drinker and the mental health of all family members.

suffer psychologically, emotionally, and socially as a result of their experience in the alcoholic environment. They are forced to play roles and meet parental needs that children in other families do not." Many develop problems in school, drug or alcohol abuse problems, depression, low self-esteem, aggressive behavior, or problems with the law. The report concludes, "Studies indicate that this population is at a very high risk for a variety of problems that could bring them to the attention of the juvenile justice system."[27]

In some towns, parents are now being held responsible for their children's criminal behavior. Though this is a controversial move, it is gathering support as people grow increasingly angry at the rise of violent juvenile crime. Parents of young criminals are being forced to pay fines and sometimes even serve jail time for their children's behavior. In those cases where the parent also has a drinking problem, a judge may see the alcohol abuse as a factor in the child's behavior, and require the parent to seek treatment.

Abusive drinking and the law has always been a difficult area in a society founded on personal freedoms. With the exception of drunk driving, the courts tend to stay away from making decisions about personal drinking habits until a serious alcohol-related crime is committed. However, attitudes toward alcohol abuse have been changing as Americans have come to recognize that it does not affect only the drinker. In recent years, much has been discovered about the link between alcohol and hidden types of crime like domestic abuse and date rape, as well as about the terrible consequences of alcohol-related disease. These discoveries have led to a less tolerant view of abusive drinking and a move toward stricter laws and penalties that cover more than drunk driving.

At the same time, the recognition of alcoholism as a disease has shifted the focus from simply punishing abusers to helping them recover from their addiction. Most people acknowledge that putting an alcoholic in jail for a few months for a drinking-related crime will not do much good unless that person also receives alcohol counseling and support in recovering from his or her illness. This combination of less tolerance with more assistance in recovery has led to many programs that use tough measures that force people to accept help for their problems and take responsibility for their lives.

Chapter 4

Do Restrictions on Alcohol Advertising Intrude on People's Lives and Livelihoods?

ANHEUSER-BUSCH, THE MAKER OF Budweiser and other popular American beers, recently introduced a program called "Bud Rewards" to promote one of its leading products. This promotion offered duffel bags, beer mugs, T-shirts, caps, and other items with the Budweiser logo in exchange for "points" earned by buying the beer.

While few people argue against a manufacturer's right to promote its products, this type of advertising has drawn a good deal of criticism. Cheryl Jones, the North Carolina chapter president of Mothers Against Drunk Driving, published these comments:

> This advertising technique is unethical for several reasons. First, it clearly rewards alcohol abuse. In order to receive "Bud Rewards," a consumer must drink substantial amounts of beer. The more beer consumed, the higher amount of points earned (one point equals one beer). For instance, loyal Bud drinkers can win a Budweiser hat after accumulating 120 points, or drinking 120 beers. After drinking 1,320 Budweisers, the lucky winner receives a Bud cooler. The ultimate prize, a Budweiser pool table, is available to a person or group that drinks 27,000 Budweisers.[28]

In her view, the program rewards problem drinkers and encourages alcohol abuse.

Another problem critics have with this type of promotion is that its target market is young drinkers. Many of the products "Bud Rewards" can buy, such as the pool table and foosball games, appeal to high school and college students. They claim it promotes drinking to a group that includes people under twenty-one who are old enough to legally drive but too young to legally drink—a dangerous combination.

Advertising is a particularly difficult area in a country that values free speech and free enterprise. Since alcohol is a legal product, and advertising is the most effective way of reaching consumers, the makers of beer, wine, and liquor depend on it for their businesses to survive. At the same time, alcohol is a potentially dangerous product, as well as illegal if consumed by minors, which is why alcohol advertising is subject to certain restrictions. Some people argue that these restrictions interfere with the manufacturers' right to free speech and fair marketing practices, while others feel they are not strong enough.

The liquor industry maintains that advertising and promotions are effective and legal sales tools. Critics charge that these ads often promote alcohol abuse and illegally target underage consumers.

Current Restrictions and a Voluntary Industry Agreement

The restrictions on advertising alcoholic beverages are not nearly as strong as those on another controversial product, tobacco. Beer and wine commercials still appear on television, though they cannot be shown during children's programming. The liquor industry, while not prohibited from advertising on radio and TV, voluntarily limited its advertising to the print media for most of the time television has been in existence. However, there have been recent moves by liquor manufacturers to change this policy, creating a controversy over whether broadcast ads for distilled spirits should be banned.

In June 1996, after a half-century of agreeing to keep liquor ads off the air, Seagram's decided to exert its right to advertise its products on television. Soon afterward, the national distillers association voted to end its voluntary broadcast advertising ban. This brought about strong protest, including a stern address on the subject by President Clinton:

> Liquor has no business with kids, and kids should have no business with liquor. Liquor ads on television would provide a message of encouragement to drink that young people simply don't need. Nothing good would come of it. Today, our message to the liquor industry is simple: For 50 years you have kept the ban, it is the responsible thing to do; for the sake of our parents and young people, please continue to keep that ban.[29]

Liquor Manufacturers' Efforts to Level the Playing Field

Despite the president's strong words, the liquor industry decided it had just as much right to advertise on TV as beer and wine manufacturers, and Seagram's refused to pull its ads. In response, Clinton asked the Federal Communications Commission (FCC) to study the impact of lifting the voluntary ban. The FCC chairman sought an official inquiry into the matter, but the commission did not reach an agreement to conduct one. The FCC commissioner at the time, Rachelle Chong, stated: "As the Supreme Court has recognized, truthful advertising—including liquor advertising—is entitled to

Reprinted with permission from Steve Kelley.

protection under the First Amendment. We cannot ignore this holding of the highest court in the land."[30] She noted that the commission's role is to see that no false or misleading advertising is aired, not that the products advertised are good for the public.

The Distilled Spirits Council of the United States (DISCUS), the trade association for the producers and marketers of liquor, defends its position on advertising in the following statement:

> Government should not ban or restrict truthful or non-deceptive liquor advertising. Advertising has not been shown to cause individuals to begin drinking or to abuse alcoholic beverages. Liquor advertising is extensively regulated by government and the industry. Effective brand competition and consumer choice depend on the free flow of commercial information.[31]

In an interview, Fred Meister, chief executive officer of DISCUS, added: "The fact, as reported by the seventh report of the Department of Health and Human Services to the United States Congress, is the Federal Trade Commission, who has responsibility for this, have said there is not a relationship between advertising and consumption, let alone abuse. The important issue is responsibility and avoiding the targeting of under-age youth. We have 26 provisions in our Code of Good Practice that are directed at responsible placement and responsible content."[32]

However, in late 1997 a new FCC chairman, William Kennard, announced that he would pursue an inquiry into liquor advertising on radio and television: "This is all about kids. It's about whether there is an appropriate role for the FCC to take in ensuring that underage drinkers are not exposed to distilled liquor advertising."[33]

Many people both within and outside the liquor industry question the fairness of being able to advertise beer and wine on the airwaves but not whiskey, vodka, and other distilled spirits. They note that while liquor consumption has declined 29 percent since 1980, sales of beer have nearly doubled since the 1960s. This change is almost certainly related to television advertising: U.S. breweries spend about $630 million each year to run commercials on TV. Fred Meister of DISCUS notes, "The beer industry spends over $600 million a year on TV advertising. And they've done that for decades. All the distillers are saying is, we want to exercise our First Amendment rights in a responsible fashion to advertise and compete for the 100 million consumers who do drink beverage alcohol."[34] He adds that, in 1997, the two largest beer companies spent more on TV advertising than the entire liquor industry spent on print advertising.

Further, some critics of this double standard note that, while campaigns against liquor advertising fear it promotes drinking among minors, it is beer—the most heavily advertised alcoholic product on TV—that young people are drinking. Tennessee district attorney general Clayburn L. Peeples notes:

> Yes, too many young people are starting to drink at an early age, but it's beer, not whiskey or gin, they are usually drinking, and the commercials we ought to be going after first are the irresponsible, hedonistic, youth oriented, seductive beer commercials, because it's beer in their blood that causes most of their car crashes and other alcohol tragedies. If your concern is for kids, go after beer.[35]

Because of all the criticism of the liquor industry's decision to advertise on television, the major networks have refused to run the ads. Groups like the National Coalition Against Drunk Driving ran successful campaigns urging people to write or call TV stations, write letters to the editors of newspapers, and contact their state and local representatives to protest broadcast liquor advertising. Laws have been proposed in Congress to ban liquor ads on TV. Public opposition to liquor commercials has sent a clear message to TV executives and the nation's lawmakers. However, there have been no serious proposals to take ads for beer or wine off the screen.

Some people suspect the liquor companies' real motive in trying to advertise on TV is to focus public attention on all televised alcohol advertising, most of which is for beer. This might result in a ban on advertising for all types of alcohol on television, which would give the manufacturers of beer, wine, and liquor equal access to the media.

The Effects of Advertising on Minors

Most of the criticism of alcohol advertising is aimed at its appeal to teens and children. Just as the Joe Camel ads for cigarettes were attacked for luring children into smoking, ads like the Budweiser commercials featuring talking frogs have been blamed for inviting young people to drink. Retired sports figures, often idols to children and teens, are featured in many of Miller's beer commercials. Other common images that appeal to teenage boys are women in bikinis, often shown being lured to young men by the promise of beer. Other ads show men on adventures in the rugged outdoors, climbing mountains or crossing rough terrain in land rovers—images that seem to make drinking and dangerous activities a natural combination.

Sex appeal and adventure have long been a major part of advertising. They are commonly used to sell everything from cars to perfume to credit cards. Their main target is a young market who is easily convinced that if they buy a certain product, it will make them more appealing to the opposite sex and add adventure to their lives. When that product is alcohol, the results can be dangerous.

Beer advertising in particular seems aimed at a young audience. While all advertising for alcohol tries to make drinking look glamorous and fun, print ads for liquor tend to use older models and a more sophisticated, mature look. Advertising for wine, both on television and in the print media, also seeks a more sophisticated appeal. The symbols found in beer ads, however, have clear appeal for young people. As one critic notes: "If you watch television at all, you know that beer commercials saturate the medium, and you know that many beer ads are socially irresponsible, encouraging consumption and crafted in such a way that they are bound to entice teenaged boys to drink beer."[36]

Alcohol advertising has been criticized for linking alcohol consumption with sex appeal, athletic prowess, and sophistication.

Beer advertising is particularly heavy during sports programming, which is watched by a large audience of teenage boys. During the 1998 Super Bowl, twenty-three minutes of ad time were devoted to beer. Because much sports programming is shown during the day on weekends, many children are exposed to its commercials.

Efforts to Limit Ads That Appeal to Minors

Among ads that have come under sharp attack are those using Halloween themes. Because Halloween is mainly a children's holiday, many people feel that using jack-o-lanterns, black cats, and witches in ads to sell alcohol is irresponsible. A national group, Coalition on Alcohol Advertising and Family Education, began a campaign in 1993 called Hands Off Halloween to stop this advertising. In 1997, the Birmingham, Alabama, chapter wrote to local merchants asking them to stop using Halloween symbols to promote alcohol. Their letter stated, "We oppose the connection of alcohol advertising with this holiday. Although Halloween is celebrated by both adults and children, the symbols of the holiday have always been associated with an appeal to the interest level of children. Adults who want to purchase alcohol for their own parties will do so without the gimmicks that greatly appeal to kids." The U.S. surgeon general, Antonia Novello, supported the campaign, noting, "The scariest thing about Halloween this year is the possibility of increased carnage on our highways and the real specter of binge drinking by our young people."[37]

Beer advertisers have responded to public pressure in recent years by pulling their ads from certain programming. In December 1996, Anheuser-Busch announced its decision to stop broadcasting its beer ads on MTV, a network popular among teenagers and preteens. It reported it was moving its ads to another network that features music videos, VH1, whose audience is older.

There have been limited studies on how much advertising influences the decision of minors to drink. Many other factors may come into play, including parents' drinking habits, peer pressure, and images of people drinking in movies and on television shows. A boy who idolizes a hard-drinking rock star may imitate that person's behavior. The makers of alcoholic beverages point to these influences as being more important factors than their ads. They argue that what the ads do is create brand loyalty—encourage people who already drink to either stay with or, if they are drinking something else, switch to their brand of alcohol.

However, many people are convinced that advertising—especially on TV—encourages many children and teens to experiment with drinking, and opposition to advertising all types of alcohol on TV is mounting. The American Academy of Pediatrics reports that American children view nearly two thousand beer and wine commercials per year. They note, "Alcohol advertising specifically targets young people by showing the supposed advantages of drinking—more friends, greater prestige, more fun, and greater sex appeal—and suggesting that without alcoholic beverages teens cannot have fun or be popular."[38]

Surveys on Advertising

A recent survey found that more than one-third of American adults support banning beer commercials. Forty-three percent of parents favored a ban. The lead researcher in the nationwide poll observed, "They're worried their kids will think more positive thoughts when they see a (Budweiser) ad with an appealing lizard or frog."[39] Half of those surveyed said they thought liquor companies were trying to influence teens to drink through their ads.

A 1994 study in California examined the relationship between television beer advertising and children's knowledge, beliefs, and intentions about drinking. It found that an awareness of the ads was related to more favorable beliefs about drinking, greater knowledge of beer brands and slogans, and increased intentions to drink as an adult. The researchers concluded: "The findings suggested that alcohol advertising may predispose young people to drinking. As a result, efforts to prevent drinking and drinking problems among young people should give attention to countering the potential effects of alcohol advertising."[40]

In 1996, the Center for Science in the Public Interest ran a study to see how adolescents respond to television beer commercials. It focused in particular on the children of alcoholics, which they described as a population at especially high risk of alcohol dependence.

Most of the adolescents in the study believed that the ads they watched modeled heavy beer consumption and that the beer companies were targeting teenagers. They associated positive characteristics with the drinkers in the ads, while they linked negative ones with the drinkers they know in real life. They showed a preference for ads that portray fantasies—the type that have been aired a lot lately—rather than realistic ones. The researchers concluded that "this study and

Eleanor Mill. Reprinted by permission of Mill NewsArt Syndicate.

previous research demonstrate sufficiently that teenagers pay a lot of attention to and are influenced by television beer advertising."[41]

Critics who are fighting to keep liquor advertising off the airwaves fear that this industry will also try to expand into the youth market. George Hacker, the director of the Alcohol Policies Project for the Center for Science in the Public Interest, noted,

> Seagram's ads are targeting radio stations that have rock 'n' roll formats, where the target audience is officially 18 and up, which probably includes fourteen and fifteen year olds as well, . . . their ads are running in prime time, . . . they're running around the NFL football games and they're reaching potentially millions of kids if those ads were spread through network TV.[42]

With so much money at stake, and with principles of a free market on the line, the makers of beer, wine, and liquor will continue to fight for their right to advertise in the United States. Few groups take the hard line of the American Academy of Pediatrics, which recommends a ban on all alcohol advertising in all media. However, those determined to stop underage drinking are sure to keep pressuring Congress to tighten restrictions on when, where, and how these products can be marketed.

Few, if any, people consider it acceptable for alcohol advertisers to target children, and companies whose ads appear to do so deny that it is their intent.

Advertising on College Campuses

Another area that has been strongly criticized is marketing alcohol to college students. College campuses pose a special problem. With the drinking age now twenty-one in all states, the majority of college students are too young to drink. However, because some undergraduates, as well as most graduate students, are of legal drinking age, advertisers are not prohibited by law from promoting alcohol on campuses. And since drinking is so popular among college students, they would be losing a valuable market if they stopped advertising at schools.

However, because of the high rates of drunk driving accidents among college-age drinkers, as well as campus violence and other problems related to drinking, there have been strong protests against targeting students. In addition, underage drinkers are exposed to the same ads as those old enough to drink.

Guidelines for Responsible Campus Marketing

Alcohol manufacturers have bowed to some of the pressure from critics. The trade associations that represent the beer, wine, and liquor industries have set up guidelines for the responsible marketing of their products to students. All three prohibit advertising that promotes heavy drinking on campus.

The Beer Institute's advertising code states, "Beer advertising and marketing materials should not depict situations where beer is being consumed excessively, in an irresponsible way," as well as "Beer advertising and marketing activities on college and university campuses or in the college media should not portray consumption of beer as being important to education, nor shall advertising directly or indirectly degrade studying."[43] The Distilled Spirits Council of the United States prohibits any advertising on campus (except in retail establishments licensed to sell alcohol) or in college newspapers.

The college market is important to the liquor industry. Though some schools prohibit alochol promotions on campus, many allow some form of advertising as long as it adheres to strict guidelines.

The policies of colleges toward alcohol promotion vary. About one-third of schools prohibit all types of alcohol promotion or advertising, and many others have advertising restrictions. However, most schools allow off-campus bars and liquor stores to advertise in their student newspapers. Many campus journalists defend alcohol advertising on free speech grounds.

Laws and Proposals for Restrictions

The federal government has not become involved with the issue of advertising alcohol to college students. Some states, however, have passed laws designed to reduce the promotion of drinking on campus. Michigan's Liquor Control Commission, for example, bans any activities on campus designed to promote the sale or drinking of alcohol, as well as manufacturers' sponsoring of campus events. Washington, Virginia, and Utah have similar laws. States that considered restricting alcohol ads in college newspapers rejected the idea as a violation of the First Amendment right to freedom of speech.

Groups such as the National Commission on Drug-Free Schools have recommended that colleges prohibit all alcohol advertising in school newspapers, at stadiums, and at all school events. They argue that because alcohol is illegal for a large portion of the student population and it interferes with creating a healthy environment for learning, its ads have no place on campus. Such groups have had limited success in convincing schools to adopt strong advertising restrictions.

Viewpoints on alcohol advertising vary widely. Some people would like to see restrictions that are at least as strong as those on cigarette advertising, which has been banned from television and radio. They are particularly concerned with the effect of ads on children and teens, who are major TV viewers. To others, however, these restrictions violate freedom of the press and interfere with the rights of the sellers of wine, beer, and liquor to do business. In recent years, the government and the courts have tended to side with the critics of alcohol advertising and imposed greater restrictions. As long as alcohol remains a legal product, a struggle is sure to continue between those whose livings depend on its promotion and those who fight to discourage abusive drinking.

Conclusion

Balancing Rights and Responsibilities

To many Americans, the government has become too big, too powerful, and too ready to interfere with individual rights. This country, they argue, was founded on basic liberties, including the right to buy and sell products in a free market. There will always be people who abuse these freedoms, whether or not the government restricts their rights. They point to the disaster of Prohibition, the period of thirteen years when the federal government attempted to outlaw alcohol throughout the country, as an example of how government interference makes the problems it tries to solve worse.

Prohibition shut down the businesses of thousands of Americans. Government agents destroyed their products and took away their right to make a living selling what for hundreds of years had been legal. The new law turned millions of Americans into criminals, as it was a law few agreed to obey. It led to the creation of a huge underground economy run by organized crime chiefs. If Prohibition stopped some Americans from drinking, it did so at an enormous cost to others. As in the drug wars today, people lost their lives in the wars over controlling the illegal liquor supply.

Sixty-five years later, the government continues to play a strong role in the control of alcohol sales and use. Some critics feel the restrictions it places on businesses that sell alcohol interfere with the right to operate in a free market. Steep licensing fees and alcohol taxes make it impossible for many small bars and liquor stores to stay in business. Since alcohol is again a legal product, many people in the beer, wine, and liquor industries feel that they are paying unfair penalties for the minority of people who use it irresponsibly.

However, to the victims of those who abuse alcohol, the restrictions on its sale and use aren't strong enough. Though it can be purchased legally by anyone twenty-one or older, alcohol is a powerful and potentially dangerous drug. For millions of Americans, it is also addictive, and its abuse often leads to illness and death. Irresponsible drinking not only costs the lives of those who drink, it also claims those of many innocent victims. Many thousands are killed by drunk drivers each year, and many children are doomed to terrible illness and early deaths because their mothers drank heavily during pregnancy.

No one wants to see drunk drivers on the roads or children suffer from fetal alcohol syndrome, and most people agree that the government has some role to play in keeping society safe from those who abuse alcohol. Yet there is widespread disagreement on how much of a role government should play, as well as how much good its interference can really do. Many alcohol abusers continue to drink irresponsibly in spite of strict laws and penalties. Large numbers of drunk drivers remain on the road even after their licenses have been suspended. Parents who lose custody of their children because of their alcohol abuse often have more children and continue to drink. Is the answer even stronger laws and penalties?

Many people think so, and their voices are being heard in states all across the country. There has been a clear trend in recent years toward increasing restrictions on the advertising, sale, and use of alcohol. After its unfortunate experiment with Prohibition, the federal government has stepped back and let state and local governments make most of their own decisions about how to control alcohol use and punish abuse. One by one, under strong pressure from groups concerned about drunk driving, the states raised their legal drinking ages to twenty-one, and many have adopted tough laws against drunk driving. Fines and sentences for drunk driving offenses are increasing, and judges are being pressured into imposing maximum sentences on drunk drivers who kill. The penalties for selling or serving alcohol to minors have increased, and the laws are being enforced more vigorously.

Even those in the business of selling alcoholic beverages have been recognizing that it is in their long-term interest to promote

At a rally celebrating the organization's tenth anniversary, members of MADD scan the photos of victims of drunk drivers. MADD is just one outlet for those who wish to promote responsible drinking.

responsible drinking. Alcohol abuse gives their products a bad name and provokes groups like MADD to press for even stronger restrictions on their sale. The manufacturers of beer, wine, and liquor are adding "responsible drinking" messages to their advertising, particularly around holidays when people tend to drink heavily, and pulling back some of their strongly criticized ads. No one wants to appear to be marketing alcohol to children or promoting heavy drinking.

However, responsible drinking is itself a complicated issue. It is now widely believed that many people are unable to drink responsibly. People with alcohol addictions are unable to control their drinking to an acceptable level. Most experts believe that, as with drug addicts, the only solution to their problem is to stop drinking. However, unlike other drugs, alcohol is a legal product. Though it may be harmful, dangerous, and sometimes even fatal, it is not against the law for alcoholics to drink. Nor have pregnant women, who can cause terrible damage to their unborn children through heavy alcohol use, been prevented from drinking. No major politician has come forward with a proposal to ban certain groups (other than those under twenty-one) from using alcohol.

In the absence of another Prohibition, which is highly unlikely, the best solutions to alcohol abuse seem to be education to prevent it, laws and penalties to try to contain it, and treatment when these measures fail. Few argue against any of these. The area where people can't seem to agree is who is responsible for controlling this powerful legal drug.

Those who feel responsibility lies with the individual, the family, and the local community don't want to see the federal government involved. Big government, they feel, is too removed from local concerns to understand them and only creates wasteful programs that interfere with individual rights. Such people would like to see less restriction on the selling of alcohol, lower taxes, and more personal responsibility, with local government stepping in as needed to address a community's problems.

To others, family and community have failed miserably in addressing problems like alcohol abuse. They point to the millions of lives lost and families destroyed by irresponsible drinking and consider the problem a national one. Such people are looking to the federal government to create nationwide laws and sentences to control alcohol-related crime. Many would also like to see higher taxes on alcohol go toward efforts to prevent and treat all forms of alcohol abuse. This group has had a stronger influence in recent years, led by powerful forces like MADD. People who have lost family members and friends to alcohol-related crime have exerted strong pressure on politicians and the courts to take strong measures against irresponsible drinking and its consequences. As long as lives continue to be destroyed by alcohol abuse, their efforts promise to continue.

NOTES

Chapter 1: Should the Sale and Use of Alcohol Be More Strictly Controlled?

1. Quoted in Doug Johnson, "New Mexico Pushes Ban on Drive-Up Liquor Sales," *Detroit News,* February 2, 1998 (http://www.detnews.com/1998/nation/9802/02/02020119.htm).
2. Quoted in Johnson, "New Mexico Pushes Ban on Drive-Up Liquor Sales."
3. Susan and Daniel Cohen, *A Six-Pack and a Fake I.D.: Teens Look at the Drinking Question.* New York: M. Evans, 1986, p. 104.
4. Quoted in Linda V. Mapes, "Arenas Seek More Liberal Rules on Hard Liquor," *Seattle Times*, January 17, 1998 (http://www.seattletimes.com/news/local/html98/booz_011798.html).
5. Quoted in Mapes, "Arenas Seek More Liberal Rules on Hard Liquor."
6. Quoted in Mapes, "Arenas Seek More Liberal Rules on Hard Liquor."
7. Quoted in Bill Callahan, "School Bus Driver in DUI Case Faces Felony Trial," *San Diego Union-Tribune*, April 17, 1998, p. B-2.

Chapter 2: Are Drunk Driving Laws Too Lenient?

8. Jon Scott Fox, DUI Defense website (http://www.duidefense.com/welcome/welcome.htm).
9. National Highway Transportation Safety Administration, "Compendium of Traffic Research Projects 1985–1995 and Beyond" (http://www.nhtsa.dot.gov/people/injury/research/COMPEND.HTM).

10. *New York Times* Editorial Desk, "One Nation, Drunk or Sober," *New York Times*, February 26, 1998.
11. Jim Battin, "End the Deadly Cycle," press release, February 1997 (http://www.assembly.ca.gov/repwww/member/280/press/OP0287.htm).
12. Quoted in National Drug Strategy Network News Brief, "National Academy of Sciences Unit Suggests Harsh New Drunk Driving Laws," April 1995 (http://www/ndsn.org/APRIL95/ALCOHOL.html).
13. *Mannes v. Gillespie*, 967 F. 2d 1310 (9th Cir. 1992), Findlaw Internet Legal Resources (http://laws.findlaw.com/9th/2/967/1310.html).
14. James B. Bakelar and Lester Grinspoon, *Drug Control in a Free Society*. Cambridge, England: Cambridge University Press, 1984, p. 137.
15. Quoted in Kathy Katella, "Drunken Drivers in the Emergency Room," *New York Times,* August 7, 1997.
16. Quoted in Matthew L. Wald, "Tough Action on Drunken Driving Pays Off," *New York Times,* November 26, 1997.

Chapter 3: Should the Law Punish Problem Drinkers?

17. Quoted in Barbara Walsh, "Drinking Becomes Partner in Crime," in "The Deadliest Drug: Maine's Addiction to Alcohol," Gannett Communications, October 24, 1997 (http://www.portland.com/alcohol/d6crime.htm).
18. Quoted in Walsh, "Drinking Becomes Partner in Crime."
19. Jack H. Mendelson and Nancy K. Mello, *Alcohol: Use and Abuse in America.* Boston: Little, Brown, 1985, p. 186.
20. Higher Education Center for Alcohol and Other Drug Prevention, "Interpersonal Violence and Alcohol and Other Drug Use," 1997 (http://pogo.edc.org/hec/pubs/factsheets/fact_sheet4.html.)
21. Minnesota Institute of Public Health, "Alcohol Use and Violence: Minnesota's Big Picture," May 1995 (http://www.miph.org/ft1.html).
22. Quoted in Walsh, "Drinking Becomes Partner in Crime."
23. Quoted in Barbara Walsh, "Police Find Drunkenness Fans

Flames of Domestic Violence," in "The Deadliest Drug: Maine's Addiction to Alcohol," Gannett Communications, October 24, 1997 (http://www.portland.com/alcohol/d6dom.htm).

24. Quoted in Carol D. Foster, Alison Landes, and Betsie B. Caldwell, *Illegal Drugs and Alcohol: America's Anguish.* Wylie, TX: Information Plus, 1991, p. 29.
25. Quoted in R. M. Morse and D. K. Flavin, "The Definition of Alcoholism," *Journal of the American Medical Association,* August 26, 1992 (http://www.thriveonline.com/health/Library/CAD/abstract6059.html).
26. Quoted in Barbara Walsh, "When Parent Drinks, Children Can Suffer Nightmarish Neglect," in "The Deadliest Drug: Maine's Addiction to Alcohol," Gannett Communications, October 24, 1997 (http://www.portland.com/alcohol/d6family.htm).
27. Johnny E. McGaha, Jack L. Stokes, and Jacqueline Melson, "Summary of Children of Alcoholism: Implications for Juvenile Justice," *Juvenile and Family Court Journal*, vol. 41, 1990 (http://www.tyc.state.tx.us/prevention/mcgaha.htm).

Chapter 4: Do Restrictions on Alcohol Advertising Intrude on People's Lives and Livelihoods?

28. Cheryl Jones, "Mothers Against Drunk Driving Voices Concerns About Anheuser-Busch 'Frequent Drinker' Program," *MADD Newsletter*, March 23, 1998 (http://www.pathfinder.com/money/latest/press/PW/1998Mar24/626.htm).
29. "Remarks by the President on Distilled Liquor Advertising," White House press release, April 1, 1997 (http://202.253.106.65:10080/pub/w.house/0403-8.txt:@=ftp.fedworld.gov).
30. "Comments of Commissioner Rachelle Chong at FCC Agenda Meeting," press release, July 9, 1997 (http://www.fcc.gov/Speeches/Chong/sprbc708.txt).
31. Distilled Spirits Council of the United States, "Liquor Advertising Restrictions," fact sheet, March 26, 1992 (http://www.discus.health.org/liqadv.htm).
32. Quoted in "Repealing Ad Prohibition," *Online News Hour*, transcript of PBS broadcast, November 12, 1996 (http://www.pbs.org/newshour/bb/business/november96/liquor_11-12.html).

33. Quoted in Coalition for the Prevention of Alcohol Problems, Action Alert, "New FCC Chairman Poised to Act on Alcohol Advertising Inquiry. Citizen Support Crucial to Countering Industry Pressure," November 4, 1997 (http://www.cspinet.org/booze/newfcc.htm).
34. Quoted in Bruce Horovitz, "Liquor Industry's Hard Line on TV Ads," *USA Today*, April 21, 1998 (http://www.usatoday.com/company/mcom038.htm).
35. Clayburn L. Peeples, "Off The Record with District Attorney General Clayburn L. Peeples: The Bitter Truth About Beer," April 7, 1997 (http://www.da28.com/columns/whiskey.htm).
36. Peeples, "Off The Record with District Attorney General Clayburn L. Peeples."
37. Quoted in Douglas Ilka, "Birmingham Group Upset by Halloween Beer Ads," *Detroit News*, October 8, 1997 (http://detnews.com/1997/metro/9710/08/10080097.htm).
38. American Academy of Pediatrics, "Children, Adolescents, and Advertising," *Pediatrics*, vol. 95, no. 2, February 1995, pp. 295–97 (http://www.aap.org/policy/00656.html).
39. Quoted in Karen Schulz, "Survey Shows Support for Giving Beer the Can," *Michigan Live*, January 23, 1998 (http://www.mlive.com/la/news/0123drytv.html).
40. J. W. Grube and L. Wallack, "Television Beer Advertising and Drinking: Knowledge, Beliefs, and Intentions Among Schoolchildren," *American Journal of Public Health*, vol. 84, no. 2, 1994, pp. 254–59.
41. Center for Science in the Public Interest Alcohol Policies Project, "Adolescent Responses to Televised Beer Advertisements: Children of Alcoholics and Others," July 1996 (http://www.espinet.org/booze/childalc.html).
42. Quoted in "Repealing Ad Prohibition."
43. Quoted in "Standards for Alcohol Marketing on Campus," in *NCADI: Last Call for High-Risk Bar Promotions That Target College Students* (http://www.health.org/pubs/lastcall/chapter3.htm).

Organizations to Contact

Against Drunk Driving (ADD)
PO Box 397 Station A, Brampton, ON L6V 2L3, CANADA
phone and fax: (905) 793-4233
e-mail: add@netcom.ca • Internet: http://www.netmediapro.com/add

Against Drunk Driving is a Canadian volunteer organization dedicated to reducing death and injury resulting from alcohol-impaired driving. ADD publishes the newsletter *Operation Lookout Network* annually, the newsletter *The ADDvisor* twice a year, and the pamphlet *The Grieving Process.*

Al-Anon Family Group Headquarters
1600 Corporate Landing Pkwy.,Virginia Beach,VA 23454
(757) 563-1600 • fax: (757) 563-1655
Internet: http://www.al/anon.alateen.org

Al-Anon is a fellowship of men, women, and children whose lives have been affected by an alcoholic family member or friend. Members share their experiences, strength, and hope to help each other and perhaps to aid in the recovery of the alcoholic. Al-Anon Family Group Headquarters provides information on its local chapters and on its affiliated organization, Alateen. Its publications include the monthly magazine the *Forum*, the semiannual *Al-Anon Speaks Out*, the bimonthly *Alateen Talk,* and several books, including *How Al-Anon Works, Path to Recovery: Steps, Traditions, and Concepts,* and *Courage to Be Me: Living with Alcoholism.*

Alcoholics Anonymous (AA)
General Service Office
PO Box 459, Grand Central Station, New York, NY 10163
(212) 870-3400 • fax: (212) 870-3003
Internet: http://www.aa.org

Alcoholics Anonymous is an international fellowship of people who are recovering from alcoholism. Because AA's primary goal is to help alcoholics remain sober, it does not sponsor research or engage in education about alcoholism. AA does, however, publish a catalog of literature concerning the organization as well as several pamphlets, including *Is AA for You? Young People and AA*, and *A Brief Guide to Alcoholics Anonymous.*

American Council on Alcohol Problems (ACAP)
3426 Bridgeland Dr., Bridgeton, MO 63044
(314) 739-5944 • fax: (314) 739-0848

ACAP is the successor to temperance organizations such as the American Temperance League and the Anti-Saloon League. It is composed of state temperance organizations, religious bodies, and fraternal organizations that support ACAP's philosophy of abstinence from alcohol. ACAP works to restrict the availability of alcohol in the United States by controlling alcohol advertising and by educating the public about the harmfulness of alcohol. It serves as a clearinghouse for information and research materials and publishes the monthly *American Issue* for donors.

Canadian Centre on Substance Abuse/Centre canadien de lutte contre l'alcoolisme et les toxicomanies (CCSA/CCLAT)
75 Albert St., Suite 300, Ottawa ON K1P 5E7, CANADA
(800) 214-4788 • (613) 235-4048 • fax: (613) 235-8101
Internet: http://www.ccsa.ca

A Canadian clearinghouse on substance abuse, the CCSA/CCLAT works to disseminate information on the nature, extent, and consequences of substance abuse and to support and assist organizations involved in substance abuse treatment, prevention, and educational programming. The CCSA/CCLAT publishes several books, including *Canadian Profile: Alcohol, Tobacco, and Other Drugs,* as well as reports, policy documents, brochures, research papers, and the newsletter *Action News.*

Distilled Spirits Council of the United States (DISCUS)
1250 I St. NW, Suite 900, Washington, DC 20005
(202) 628-3544

The Distilled Spirits Council of the United States is the national trade association representing producers and marketers of distilled spirits in the United States. It seeks to ensure the responsible advertising and marketing of distilled spirits to adult consumers and to prevent such advertising and marketing from targeting individuals below the legal purchase age. DISCUS publishes fact sheets, the periodic newsletter *News Release*, and several pamphlets, including *The Drunk Driving Prevention Act.*

Entertainment Industries Council (EIC)
1760 Reston Pkwy., Suite 415, Reston, VA 20190-3303
(703) 481-4404 • fax: (703) 481-1418
e-mail: East Coast: eiceast@aol.com • West Coast: eicwest@aol.com

The EIC works to educate the entertainment industry and their audiences about major public health and social issues. Its members strive to effect social change by providing educational materials, research, and training to the entertainment industry. The EIC publishes several fact sheets on alcohol, children of alcoholics, women and addiction, and alcohol-impaired driving.

Mothers Against Drunk Driving (MADD)
511 E. John Carpenter Frwy., No. 700, Irving, TX 75062
(800) GET-MADD (438-6233)
e-mail: Information: info@madd.org • Victim's Assistance:
victims@madd.org • Internet: http://www.madd.org

Mothers Against Drunk Driving seeks to act as the voice of victims of drunk driving accidents by speaking on their behalf to communities, businesses, and educational groups and by providing materials for use in medical facilities and health and driver education programs. MADD publishes the biannual *MADDvocate for Victims Magazine* and the newsletter *MADD in Action* as well as a variety of brochures and other materials on drunk driving.

National Council on Alcoholism and Drug Dependence (NCADD)
12 W. 21st St., New York, NY 10010
(212) 206-6770 • fax: (212) 645-1690
Internet: http: //www.ncadd.org

NCADD is a volunteer health organization that helps individuals overcome addictions, advises the federal government on drug and alcohol policies, and develops substance abuse prevention and education programs for youth. It publishes fact sheets, such as *Youth and Alcohol,* and pamphlets, such as *Who's Got the Power? You . . . or Drugs?*

Office for Substance Abuse Prevention (OSAP)
National Clearinghouse for Alcohol and Drug Information (NCADI)
PO Box 2345, Rockville, MD 20847-2345
(800) 729-6686 • (301) 468-2600
TDD: (800) 487-4889 or (301) 230-2867
Internet: http://www.health.org

OSAP leads U.S. government efforts to prevent alcoholism and other drug problems among Americans. Through the NCADI, OSAP provides the public with a wide variety of information on alcoholism and other addictions. Its publications include the bimonthly *Prevention Pipeline,* the fact sheet *Alcohol Alert*, monographs such as "Social Marketing/Media Advocacy" and "Advertising and Alcohol," brochures, pamphlets, videotapes, and posters. Publications in Spanish are also available.

Rational Recovery Systems (RRS)
PO Box 800, Lotus, CA 95651
(916) 621-4374 • (800) 303-CURE • phone and fax (916) 621-2667
e-mail: Self-Help Network: rrsn@rational.org • Training Institute: rrti@rational.org • Centers: rrc@rational.org • Political and Legal Action Network: rrplan@rational.org
Internet: http://www.rational.org/recovery

RRS is a national self-help organization that offers a cognitive rather than spiritual approach to recovery from alcoholism. Its philosophy holds that alcoholics can attain sobriety without depending on other people or a "higher power." Rational Recovery Systems publishes materials about the organization and its use of rational-emotive therapy.

Research Society on Alcoholism (RSA)
4314 Medical Pkwy., No. 300, Austin, TX 78756
phone and fax: (512) 454-0022
e-mail: debbyrsa@bga.com

The RSA provides a forum for researchers who share common interests in alcoholism. The society's purpose is to promote research on the prevention and treatment of alcoholism. It publishes the journal *Alcoholism: Clinical and Experimental Research* nine times a year as well as the book series Recent Advances in Alcoholism.

Secular Organizations for Sobriety (SOS)
PO Box 5, Buffalo, NY 14215
(716) 834-2922

SOS is a network of groups dedicated to helping individuals achieve and maintain sobriety. The organization believes that alcoholics can best recover by rationally choosing to make sobriety rather than alcohol a priority. Most members of SOS reject the religious basis of Alcoholics Anonymous and other similar self-help groups. SOS publishes the quarterly *SOS International Newsletter* and distributes the books *Unhooked: Staying Sober and Drug Free* and *How to Stay Sober: Recovery Without Religion*, written by SOS founder James Christopher.

Students Against Drunk Driving (SADD)
PO Box 800, Marlboro, MA 01752
(508) 481-3568

SADD offers help to schools that want to set up their own chapters. It encourages peer counseling among students on saying no to drinking and drugs and aims to increase public awareness about drunk driving.

For Further Reading

Edward Behr, *Prohibition: Thirteen Years That Changed America.* New York: Arena, 1996. This history of Prohibition describes the times that led to the ban on alcohol, the bootleggers and gangsters who flourished during the period, and the politicians who openly encouraged illegal alcohol sales.

Eve Bunting, *A Sudden Silence.* New York: Fawcett Books, 1989. This novel for young adults describes the feelings and experiences of a boy whose brother was killed by a hit-and-run driver. It follows his efforts to track down the killer and discusses the problem of alcoholism and how society deals with it.

Michael Dorris, *The Broken Cord: A Family's Ongoing Struggle with Fetal Alcohol Syndrome.* New York: Harper & Row, 1989. The author describes his experiences raising an adopted son who suffered from fetal alcohol syndrome. The boy's mother had died of alcohol poisoning, and he grew up suffering from intelligence, judgment, and physical problems, as well as alcohol addiction. Also discussed is the particularly serious problem of alcoholism among native Americans.

Jack Erdman with Larry Kearney, *Whiskey's Children: An Inspiring True Story of Struggle and Redemption.* New York: Kensington Books, 1994. In this autobiography of a fourth-generation alcoholic, the author describes his journey from the pain and despair of addiction to recovery.

Peter Grier, "Drunk Driving Draws Global Wrath," *Christian Science Monitor,* September 3, 1997. (http://www.csmonitor.com/durable/1997/09/03/us/us.1.html). This article discusses drunk driving as an international problem and compares American attitudes and laws with those abroad.

Jerome D. Levin, *Recovery from Alcoholism.* Northvale, NJ: Jason Aronson, 1991. Through case histories, this book illustrates how alcohol affects addicted drinkers and discusses approaches to treatment.

Eric Pianin, "How Pressure Politics Bottled Up a Tough Drunk-Driving Rule," *Washington Post,* May 22, 1998. (http://www.washingtonpost.com/wp-srv/Wplate/1998-05/22/0931-052298-idx.html). This article describes a recent effort by sixty-two senators and President Clinton to enact a federal drunk driving law that would lower the BAC limit to .08 in all states, and discusses how the alcoholic beverage and restaurant industries pressured members of Congress into rejecting it.

Works Consulted

Books

James B. Bakelar and Lester Grinspoon, *Drug Control in a Free Society.* Cambridge, England: Cambridge University Press, 1984. This book surveys the history of alcohol use and regulation in the United States and examines how the law has affected use and abuse over the years.

Susan and Daniel Cohen, *A Six-Pack and a Fake I.D.: Teens Look at the Drinking Question.* New York: M. Evans, 1986. The authors discuss how raising the legal drinking age to twenty-one nationwide has affected teen attitudes and behavior.

Carol D. Foster, Alison Landes, and Betsie B. Caldwell, *Illegal Drugs and Alcohol: America's Anguish.* Wylie, TX: Information Plus, 1991. This book covers a wide range of issues on alcohol use, including its effects on the body and mind, alcoholism, drunk driving, and the connection between alcohol abuse and violence.

Jack H. Mendelson and Nancy K. Mello, *Alcohol: Use and Abuse in America.* Boston: Little, Brown, 1985. The authors provide a comprehensive history of alcohol use in the United States, from the time of the first settlers through the modern era.

Periodicals

Bill Callahan, "School Bus Driver in DUI Case Faces Felony Trial," *San Diego Union-Tribune,* April 17, 1998.

Cindy Glover, "Bill Allows Votes on Drive-Up Liquor Bans," *ABQ Journal*, January 22, 1998. (http://www.abqjournal.com/news/xgr98/2legisl-22.htm)

J. W. Grube and L. Wallack, "Television Beer Advertising and Drinking: Knowledge, Beliefs, and Intentions Among School-children," *American Journal of Public Health,* vol. 84, no. 2, 1994.

Bruce Horovitz, "Liquor Industry's Hard Line on TV Ads," *USA Today,* April 21, 1998. (http://www.usatoday.com/company/mcom038.htm)

Douglas Ilka, "Birmingham Group Upset by Halloween Beer Ads," *Detroit News,* October 8, 1997. (http://detnews.com/1997/metro/9710/08/10080097.htm)

Doug Johnson, "New Mexico Pushes Ban on Drive-Up Liquor Sales," *Detroit News,* February 2, 1998. (http://www.detnews.com/1998/nation/9802/02/02020119.htm)

Kathy Katella, "Drunken Drivers in the Emergency Room," *New York Times,* August 7, 1997.

Linda V. Mapes, "Arenas Seek More Liberal Rules on Hard Liquor," *Seattle Times,* January 17, 1998. (http://www.seattletimes.com/news/local/html98/booz_011798.html)

Johnny E. McGaha, Jack L. Stokes, and Jacqueline Melson, "Summary of Children of Alcoholism: Implications for Juvenile Justice," *Juvenile and Family Court Journal*, vol. 41, 1990. (http://www.tyc.state.tx.us/prevention/mcgaha.htm)

Bob Miller, "Drunken Driver Convicted of Murder," *Washington Post,* April 16, 1998. (http://search.washingtonpost.com/wp-srv/WAPO/19980416/V000263-041698-idx.html)

R. M. Morse and D. K. Flavin, "The Definition of Alcoholism," *Journal of the American Medical Association,* August 26, 1992. (http://www.thriveonline.com/health/Library/CAD/abstract6059.html)

New York Times Editorial Desk, "One Nation, Drunk or Sober," *New York Times,* February 26, 1998.

Susan Schramm, "Victim's Family Livid After Drunk-Driving Sentence," *Indianapolis Star/News,* November 11, 1997. (http://www.starnews.com/News/indianapolis/97/nov/1111SN_drunk.html)

Karen Schulz, "Survey Shows Support for Giving Beer the Can," *Michigan Live,* January 23, 1998. (http://www.mlive.com/la/news/0123drytv.html)

Matthew L. Wald, "Tough Action on Drunken Driving Pays Off," *New York Times*, November 26, 1997.

Barbara Walsh, "Drinking Becomes Partner in Crime," in "The Deadliest Drug: Maine's Addiction to Alcohol," Gannett Communications, October 24, 1997. (http://www.portland.com/alcohol/d6crime.htm)

Barbara Walsh, "Police Find Drunkenness Fans Flames of Domestic Violence," in "The Deadliest Drug: Maine's Addiction to Alcohol," Gannett Communications, October 24, 1997. (http://www.portland.com/alcohol/d6dom.htm)

Barbara Walsh, "When Parent Drinks, Children Can Suffer Nightmarish Neglect," in "The Deadliest Drug: Maine's Addiction to Alcohol," Gannett Communications, October 24, 1997. (http://www.portland.com/alcohol/d6family.htm)

Speeches, Interviews, and Press Releases

Jim Battin, "End the Deadly Cycle," press release, February 1997. (http://www.assembly.ca.gov/repwww/member/a80/press/op0297.htm)

"Comments of Commissioner Rachelle Chong at FCC Agenda Meeting," press release, July 9, 1997. (http://www.fcc.gov/Speeches/Chong/sprbc708.txt)

Clayburn L. Peeples, "Off the Record with District Attorney General Clayburn L. Peeples: The Bitter Truth About Beer," April 7, 1997. (http://www.da28.com/columns/whiskey.htm)

"Remarks by the President on Distilled Liquor Advertising," White House press release, April 1, 1997. (http://202.253.106.65:10080/pub/w.house/0403-8.txt:@=ftp.fedworld.gov)

"Repealing Ad Prohibition," *Online News Hour,* transcript of PBS broadcast, November 12, 1996. (http://www.pbs.org/newshour/bb/business/november96/liquor_11-12.html)

U.S. Department of Justice, Bureau of Justice Statistics, "Four in Ten Criminal Offenders Report Alcohol as a Factor in Violence," press release, April 5, 1998. (http://www.health.org/pressrel/april98/2.htm)

Newsletters and Other Organization Publications

AAA Michigan, News and Information, "Tough Drunk Driving Laws Will Take Effect April 1." (http://www.aaamich.com/information/newswire/dui/html)

American Academy of Pediatrics, "Children, Adolescents, and Advertising," *Pediatrics,* vol. 95, no. 2, February 1995. (http://www.aap.org/policy/00656.html)

California State Automobile Association, Education and Safety, "A DUI Arrest: What Happens?" (http://www.csaa.com/education/smart/dui-edusafe)

Center for Science in the Public Interest Alcohol Policies Project, "Adolescent Responses to Televised Beer Advertisements: Children of Alcoholics and Others," July 1996. (http://www.espinet.org/booze/childalc.html)

Coalition for the Prevention of Alcohol Problems, Action Alert, "New FCC Chairman Poised to Act on Alcohol Advertising Inquiry. Citizen Support Crucial to Countering Industry Pressure," November 4, 1997. (http://www.cspinet.org/booze/newfcc.htm)

Department of the Treasury, Bureau of Alcohol, Tobacco, and Firearms, "Liquor Laws and Regulations for Retail Dealers," August 1995. (http://www.atf.treas.gov/pub/5170-2a.htm)

Distilled Spirits Council of the United States, "Liquor Advertising Restrictions," fact sheet, March 26, 1992. (http://www.discus.health.org/liqadv.htm)

Distilled Spirits Council of the United States, "Tax Increases on Distilled Spirits Don't Work," fact sheet, (http://www.discus.health.org/factsheet.htm)

Jon Scott Fox, DUI defense website. (http://www.duidefense.com/welcome/welcome.htm)

Higher Education Center for Alcohol and Other Drug Prevention, "Interpersonal Violence and Alcohol and Other Drug Use," 1997. (http://pogo.edc.org/hec/pubs/factsheets/fact_sheet4.html)

Cheryl Jones, "Mothers Against Drunk Driving Voices Concerns About Anheuser-Busch 'Frequent Drinker' Program," *MADD Newsletter,* March 23, 1998. (http://www.pathfinder.com/money/latest/press/PW/1998Mar24/626.htm)

Libertarian Party, "TV Liquor Ban Is Attack on Free Speech and Trade," news release, April 3, 1997. (http://www.lp.org/rel/970403-liquor.html)

MADD Website, home page, "About MADD." (http://madd.org/aboutmadd/default.shtml)

Michigan Beer and Wine Wholesalers Association, "Beer/Wine by Mail Controversial." (http://www.mbwwa.org/whatshot.html)

Minnesota Institute of Public Health, "Alcohol Use and Violence: Minnesota's Big Picture," May 1995. (http://www.miph.org/ft1.html)

National Commission Against Drunk Driving, "Alcohol and Alcohol-Related Problems: A Sobering Look." (http://www.ncadd.org/problem.htm)

National Drug Strategy Network News Brief, "National Academy of Sciences Unit Suggests Harsh New Drunk Driving Laws," April 1995. (http://www.ndsn.org/APRIL95/ALCOHOL.html)

National Highway Transportation Safety Administration, "Compendium of Traffic Research Projects 1985–1995 and Beyond." (http://www.nhtsa.dot.gov/people/injury/research/COMPEND.HTM)

Racers Against Drunk Driving, "1995 Summary of Statistics: The Impaired Driving Program." (http://www.raddracing.com/news/radd4.html)

"Standards for Alcohol Marketing on Campus," in *NCADI: Last Call for High-Risk Bar Promotions That Target College Students*. (http://www.health.org/pubs/lastcall/chapter3.htm)

INDEX

Picture Credits

Cover photo: © David Young Wolff/Tony Stone Images

© copyright 1988 PhotoDisc, Inc., 12, 20, 28, 32, 34, 39, 44, 47, 55 (bottom)

FPG International, Inc., 23

© Robert Fox/Impact Visuals, 9

© Bernard Gotfryd/Woodfin Camp & Associates, Inc., 55 (top)

© Jack Kurtz/Impact Visuals, 63

Library of Congress, 8

Sue Ogrocki/SIPA Press, 51

Reuters/Corbis-Bettmann, 19

© Loren Santow/Impact Visuals, 58

UPI/Corbis-Bettmann, 27, 71

© Piet van Lier/Impact Visuals, 67

© Jim West/Impact Visuals, 18

About the Author

Lisa Wolff is a writer and editor with many years of staff experience at New York publishing houses. She currently lives in San Diego, where she edits reference books and writes articles on health and books on social issues.